D1087652

Solved and Unsolved Problems in Number Theory

Solved and Unsolved Problems in Number Theory

By DANIEL SHANKS

CHELSEA PUBLISHING COMPANY
NEW YORK, N.Y.

SECOND EDITION

Library of Congress Cataloging in Publication Data

Shanks, Daniel.
    Solved and unsolved problems in number theory.

    Bibliography: p.
    Includes index.
    1. Numbers, Theory of. I. Title.
[QA241.S44 1978] 512'.7 77-13010
ISBN 0-8284-0297-3

Printed on 'long-life' acid-free paper

Printed in the United States of America

CONTENTS

Chapter III

PYTHAGOREANISM AND ITS MANY CONSEQUENCES

PREFACE TO THE SECOND EDITION

The Preface to the First Edition (1962) states that this is "a rather tightly organized presentation of elementary number theory" and that "number theory is very much a live subject." These two facts are in conflict fifteen years later. Considerable updating is desirable at many places in the 1962 text, but the needed insertions would call for drastic surgery. This could easily damage the flow of ideas and the author was reluctant to do that. Instead, the original text has been left as is, except for typographical corrections, and a brief new chapter entitled "Progress" has been added. A new reader will read the book at two levels—as it was in 1962, and as things are today.

Of course, not all advances in number theory are discussed, only those pertinent to the earlier text. Even then, the reader will be impressed with the changes that have occurred and will come to believe—if he did not already know it—that number theory is very much a live subject.

The new chapter is rather different in style, since few topics are developed at much length. Frequently, it is extremely brief and merely gives references. The intent is not only to discuss the most important changes in sufficient detail but also to be a useful guide to many other topics. A propos this intended utility, one special feature: Developments in the algorithmic and computational aspects of the subject have been especially active. It happens that the author was an editor of *Mathematics of Computation* throughout this period, and so he was particularly close to most of these developments. Many good students and professionals hardly know this material at all. The author feels an obligation to make it better known, and therefore there is frequent emphasis on these aspects of the subject.

To compensate for the extreme brevity in some topics, numerous references have been included to the author's own reviews on these topics. They are intended especially for any reader who feels that he must have a second helping. Many new references are listed, but the following economy has been adopted: if a paper has a good bibliography, the author has usually refrained from citing the references contained in that bibliography.

The author is grateful to friends who read some or all of the new chapter. Especially useful comments have come from Paul Bateman, Samuel Wagstaff, John Brillhart, and Lawrence Washington.

DANIEL SHANKS
December 1977

PREFACE TO THE FIRST EDITION

It may be thought that the title of this book is not well chosen since the book is, in fact, a rather tightly organized presentation of elementary number theory, while the title may suggest a loosely organized collection of problems. Nonetheless the nature of the exposition and the choice of topics to be included or omitted are such as to make the title appropriate. Since a preface is the proper place for such discussion we wish to clarify this matter here.

Much of elementary number theory arose out of the investigation of three problems; that of perfect numbers, that of periodic decimals, and that of Pythagorean numbers. We have accordingly organized the book into three long chapters. The result of such an organization is that *motivation* is stressed to a rather unusual degree. Theorems arise in response to previously posed problems, and their proof is sometimes delayed until an appropriate analysis can be developed. These theorems, then, or most of them, are "solved problems." Some other topics, which are often taken up in elementary texts—and often dropped soon after—do not fit directly into these main lines of development, and are postponed until Volume II. Since number theory is so extensive, *some* choice of topics is essential, and while a common criterion used is the personal preferences or accomplishments of an author, there is available this other procedure of following, rather closely, a few main themes and postponing other topics until they become necessary.

Historical discussion is, of course, natural in such a presentation. However, our primary interest is in the theorems, and their logical interrelations, and not in the history *per se*. The aspect of the historical approach which mainly concerns us is the determination of the problems which suggested the theorems, and the study of which provided the concepts and the techniques which were later used in their proof. In most number theory books *residue classes* are introduced prior to Fermat's Theorem and the Reciprocity Law. But this is not at all the correct historical order. We have here restored these topics to their historical order, and it seems to us that this restoration presents matters in a more natural light.

The "unsolved problems" are the conjectures and the open questions—we distinguish these two categories—and these problems are treated more fully than is usually the case. The conjectures, like the theorems, are introduced at the point at which they arise naturally, are numbered and stated formally. Their significance, their interrelations, and the heuristic evidence supporting them are often discussed. It is well known that some unsolved problems, such as Fermat's Last Theorem and Riemann's Hypothesis, have been enormously fruitful in suggesting new mathematical fields, and for this reason alone it is not desirable to dismiss conjectures without an adequate discussion. Further, number theory is very much a live subject, and it seems desirable to emphasize this.

So much for the title. The book is largely an exposition of known and fundamental results, but we have included several original topics such as cycle graphs and the circular parity switch. Another point which we might mention is a tendency here to analyze and mull over the proofs—to study their strategy, their logical interrelations, their possible simplifications, etc. It happens that such considerations are of particular interest to the author, and there may be some readers for whom the theory of proof is as interesting as the theory of numbers. However, for all readers, such analyses of the proofs should help to create a deeper understanding of the subject. That is their main purpose. The historical introductions, especially to Chapter III, may be thought by some to be too long, or even inappropriate. We need not contest this, and if the reader finds them not to his taste he may skip them without much loss.

The notes upon which this book was based were used as a text at the American University during the last year. A three hour first course in number theory used the notes through Sect. 48, omitting the historical Sects. 41–45. But this is quite a bit of material, and another lecturer may prefer to proceed more slowly. A second semester, which was partly lecture and partly seminar, used the rest of the book and part of the forthcoming Volume II. This included a proof of the Prime Number Theorem and would not be appropriate in a first course.

The exercises, with some exceptions, are an integral part of the book. They sometimes lead to the next topic, or hint at later developments, and are often referred to in the text. Not every reader, however, will wish to work every exercise, and it should be stated that while some are very easy, others are not. The reader should not be discouraged if he cannot do them all. We would ask, though, that he read them, even if he does not do them.

The book was not written solely as a textbook, but was also meant for the technical reader who wishes to pursue the subject independently. It is a somewhat surprising fact that although one never meets a mathematician who will say that he doesn't know calculus, algebra, etc., it is quite common to have one say that he doesn't know any number theory. Yet this is an old, distinguished, and highly praised branch of mathematics, with contributions on the highest level, Gauss, Euler, Lagrange, Hilbert, etc. One might hope to overcome this common situation by a presentation of the subject with sufficient motivation, history, and logic to make it appealing.

If, as they say, we can succeed even partly in this direction we will consider ourselves well rewarded.

The original presentation of this material was in a series of twenty public lectures at the David Taylor Model Basin in the Spring of 1961. Following the precedent set there by Professor F. Murnaghan, the lectures were written, given, and distributed on a weekly schedule.

Finally, the author wishes to acknowledge, with thanks, the friendly advice of many colleagues and correspondents who read some, or all of the notes. In particular, helpful remarks were made by A. Sinkov and P. Bateman, and the author learned of the Original Legendre Symbol in a letter from D. H. Lehmer. But the author, as usual, must take responsibility for any errors in fact, argument, emphasis, or presentation.

DANIEL SHANKS
May 1962

CHAPTER I

FROM PERFECT NUMBERS TO THE QUADRATIC RECIPROCITY LAW

1. Perfect Numbers

Many of the basic theorems of number theory stem from two problems investigated by the Greeks—the problem of *perfect* numbers and that of *Pythagorean* numbers. In this chapter we will examine the former, and the many important concepts and theorems to which their investigation led. For example, the first extensive table of primes (by Cataldi) and the very important Fermat Theorem were, as we shall see, both direct consequences of these investigations. Euclid's theorems on primes and on the greatest common divisor, and Euler's theorems on quadratic residues, may also have been such consequences but here the historical evidence is not conclusive. In Chapter III we will take up the Pythagorean numbers and their many historic consequences but for now we will confine ourselves to perfect numbers.

Definition 1. A *perfect* number is equal to the sum of all its positive divisors other than itself. (Euclid.)

Example: Since the positive divisors of 6 other than itself are 1, 2, and 3 and since

$$1 + 2 + 3 = 6,$$

6 is perfect.

The first four perfect numbers, which were known to the Greeks, are

$$P_1 = 6,$$

$$P_2 = 28,$$

$$P_3 = 496,$$

$$P_4 = 8128.$$

In the Middle Ages it was asserted repeatedly that P_m, the mth perfect number, was always exactly m digits long, and that the perfect numbers alternately end in the digit 6 and the digit 8. Both assertions are false. In fact there is no perfect number of 5 digits. The next perfect number is

$$P_5 = 33{,}550{,}336.$$

Again, while this number does end in 6, the next does not end in 8. It also ends in 6 and is

$$P_6 = 8{,}589{,}869{,}056.$$

We must, therefore, at least weaken these assertions, and we do so as follows: The first we change to read

Conjecture 1. *There are infinitely many perfect numbers.*

The second assertion we split into two distinct parts:

Open Question 1. *Are there any odd perfect numbers?*

Theorem 1. *Every even perfect number ends in a* 6 *or an* 8.

By a *conjecture* we mean a proposition that has not been proven, but which is favored by some serious evidence. For Conjecture 1, the evidence is, in fact, not very compelling; we shall examine it later. But primarily we will be interested in the body of theory and technique that arose in the attempt to settle the conjecture.

An *open question* is a problem where the evidence is not very convincing one way or the other. Open Question 1 has, in fact, been "conjectured" in both directions. Descartes could see no reason why there should *not* be an odd perfect number. But none has ever been found, and there is no odd perfect number less than a trillion, if any. Hardy and Wright said there probably are no odd perfect numbers at all—but gave no serious evidence to support their statement.

A *theorem*, of course, is something that has been proved. There are important theorems and unimportant theorems. Theorem 1 is curious but not important. As we proceed we will indicate which are the important theorems.

The distinction between open question and conjecture is, it is true, somewhat subjective, and different mathematicians may form different judgments concerning a particular proposition. We trust that there will be no similar ambiguity concerning the theorems, and we shall prove many such propositions in the following pages. Further, in some instances, we shall not merely prove the theorem but also discuss the nature of the proof, its strategy, and its logical dependence upon, or independence from, some concept or some previous theorem. We shall sometimes inquire whether the proof can be simplified. And, if we state that Theorem T is particularly important, then we should explain why it is important, and how its fundamental role enters into the structure of the subsequent theorems.

Before we prove Theorem 1, let us rewrite the first four perfects in binary notation. Thus:

	Decimal	*Binary*
P_1	6	110
P_2	28	11100
P_3	496	111110000
P_4	8128	1111111000000

Now a binary number consisting of n 1's equals $1 + 2 + 4 + \cdots + 2^{n-1} = 2^n - 1$. For example, 11111 (binary) $= 2^5 - 1 = 31$ (decimal). Thus all of the above perfects are of the form

$$2^{n-1}(2^n - 1),$$

e.g.,

$$496 = 16 \cdot 31 = 2^4(2^5 - 1).$$

Three of the thirteen books of Euclid were devoted to number theory. In Book IX, Prop. 36, the final proposition in these three books, he proves, in effect,

Theorem 2. *The number* $2^{n-1}(2^n - 1)$ *is perfect if* $2^n - 1$ *is a prime.*

It appears that Euclid was the first to define a prime—and possibly in this connection. A modern version is

Definition 2. If p is an integer, >1, which is divisible only by ± 1 and by $\pm p$, it is called *prime*. An integer >1, not a prime, is called *composite.*

About 2,000 years after Euclid, Leonhard Euler proved a converse to Theorem 2:

Theorem 3. *Every even perfect number is of the form* $2^{n-1}(2^n - 1)$ *with* $2^n - 1$ *a prime.*

We will make our proof of Theorem 1 depend upon this Theorem 3 (which will be proved later), and upon a simple theorem which we shall prove at once:

Theorem 4 (Cataldi-Fermat). *If* $2^n - 1$ *is a prime, then* n *is itself a prime.*

Proof. We note that

$$a^n - 1 = (a - 1)(a^{n-1} + a^{n-2} + \cdots + a + 1).$$

If n is not a prime, write it $n = rs$ with $r > 1$ and $s > 1$. Then

$$2^n - 1 = (2^r)^s - 1,$$

and $2^n - 1$ is divisible by $2^r - 1$, which is >1 since $r > 1$.

Assuming Theorem 3, we can now prove Theorem 1.

PROOF OF THEOREM 1. If N is an even perfect number,

$$N = 2^{p-1}(2^p - 1)$$

with p a prime. Every prime >2 is of the form $4m + 1$ or $4m + 3$, since otherwise it would be divisible by 2. Assume the first case. Then

$$\begin{aligned} N &= 2^{4m}(2^{4m+1} - 1) \\ &= 16^m(2 \cdot 16^m - 1) \qquad \text{with } m \geqq 1. \end{aligned}$$

But, by induction, it is clear that 16^m always ends in 6. Therefore $2 \cdot 16^m - 1$ ends in 1 and N ends in 6. Similarly, if $p = 4m + 3$,

$$N = 4 \cdot 16^m(8 \cdot 16^m - 1)$$

and $4 \cdot 16^m$ ends in 4, while $8 \cdot 16^m - 1$ ends in 7. Thus N ends in 8. Finally if $p = 2$, we have $N = P_1 = 6$, and thus all even perfects must end in 6 or 8.

2. EUCLID

So far we have not given any insight into the reasons for $2^{p-1}(2^p - 1)$ being perfect—if $2^p - 1$ is prime. Theorem 2 would be extremely simple were it not for a rather subtle point. Why should $N = 2^{p-1}(2^p - 1)$ be perfect? The following positive integers divide N:

$$\begin{gathered} 1 \text{ and } (2^p - 1) \\ 2 \text{ and } 2(2^p - 1) \\ 2^2 \text{ and } 2^2(2^p - 1) \\ \vdots \\ 2^{p-1} \text{ and } 2^{p-1}(2^p - 1) \end{gathered}$$

Thus Σ, the sum of these divisors, *including* the last, $2^{p-1}(2^p - 1) = N$, is equal to

$$\Sigma = (1 + 2 + 2^2 + \cdots + 2^{p-1})[1 + (2^p - 1)].$$

Summing the geometric series we have

$$\Sigma = (2^p - 1) \cdot 2^p = 2N.$$

Therefore the sum of *these* divisors, but not counting N itself, is equal to $\Sigma - N = N$. Does this make N perfect? Not quite. How do we know there are no other positive divisors?

Euclid, recognizing that this needed proof, provided two fundamental underlying theorems, Theorem 5 and Theorem 6 (below), and one fundamental algorithm.

Definition 3. If g is the greatest integer that divides both of two integers, a and b, we call g their *greatest common divisor*, and write it

$$g = (a, b).$$

In particular, if

$$(a, b) = 1,$$

we say that a is *prime to* b.

EXAMPLES:

$2 = (4, 14)$	$1 = (1, n)$	(any n)
$3 = (3, 9)$	$1 = (n - 1, n)$	(any n)
$1 = (9, 20)$	$1 = (p, q)$	(any two distinct primes)

Definition 4. If a divides b, we write

$$a|b;$$

if not we write

$$a \nmid b.$$

EXAMPLE:

$$23|2047.$$

Theorem 5 (Euclid). *If $g = (a, b)$ there is a linear combination of a and b with integer coefficients m and n (positive, negative, or zero) such that*

$$g = ma + nb.$$

Assuming this theorem, which will be proved later, we easily prove a

Corollary. *If* $(a, c) = (b, c) = 1$, *then* $(ab, c) = 1$.
PROOF. We have

$$m_1a + n_1 c = 1 \quad \text{and} \quad m_2b + n_2 c = 1,$$

and therefore, by multiplying,

$$Mab + Nc = 1$$

with $M = m_1m_2$ and $N = m_1n_2a + m_2n_1b + n_1n_2c$. Then any common divisor of ab and c must divide 1, and therefore $(ab, c) = 1$.

We also easily prove

Theorem 6 (Euclid). *If a, b, and c are integers such that*

$$c|ab \quad \text{and} \quad (c, a) = 1,$$

then

$$c|b.$$

PROOF. By Theorem 5,

$$mc + na = 1.$$

Therefore

$$mcb + nab = b,$$

but since $c|ab$, $ab = cd$ for some integer d. Thus

$$c(mb + nd) = b,$$

or $c|b$.

Corollary. *If a prime p divides a product of n numbers,*

$$p|a_1a_2 \cdots a_n ,$$

it must divide at least one of them.

PROOF. If $p \nmid a_1$, then $(a_1, p) = 1$. If now, $p \nmid a_2$, then we must have $p \nmid a_1a_2$, for, by the theorem, if $p|a_1a_2$, then $p|a_2$. It follows that if $p \nmid a_1$, $p \nmid a_2$, and $p \nmid a_3$, then $p \nmid a_1a_2a_3$. By induction, if p divided none of a's it could not divide their product.

Euclid did not give Theorem 7, the *Fundamental Theorem of Arithmetic*, and it is not necessary—in this generality—for Euclid's Theorem 2. But we do need it for Theorem 3.

Theorem 7. *Every integer, N, > 1, has a unique factorization into primes, p_i in a* standard form,

$$N = p_1^{a_1}p_2^{a_2} \cdots p_n^{a_n}, \tag{1}$$

with $a_i > 0$ and $p_1 < p_2 < \cdots < p_n$. That is, if

$$N = q_1^{b_1}q_2^{b_2} \cdots q_m^{b_m} \tag{2}$$

for primes $q_1 < q_2 < \cdots < q_m$ and exponents $b_i > 0$, then $p_i = q_i$, $m = n$, and $a_i = b_i$.

PROOF. First, N must have at least one representation, Eq. (1). Let a be the *smallest* divisor of N which is >1. It must be a prime, since if not, a would have a divisor >1 and $<a$. This divisor, $<a$, would divide N and this contradicts the definition of a. Write a now as p_1, and the quotient, N/p_1, as N_1. Repeat the process with N_1. The process must terminate, since

$$N > N_1 > N_2 > \cdots > 1,$$

This generates Eq. (1). Now if there were a second representation, by the corollary of Theorem 6, each p_i must equal some q_i, since $p_i|N$. Likewise each q_i must equal some p_i. Therefore $p_i = q_i$ and $m = n$. If $b_i > a_i$, divide $p_i^{a_i}$ into Eqs. (1) and (2). Then p_i would divide the quotient in Eq. (2) but not in Eq. (1). This contradiction shows that $a_i = b_i$.

Corollary. *The only positive divisors of*

$$N = p_1^{a_1} \cdots p_n^{a_n}$$

are those of the form

$$p_1^{c_1} p_2^{c_2} \cdots p_n^{c_n} \tag{3}$$

where

$$0 \leqq c_i \leqq a_i .$$

PROOF. Let $f|N$ and write $N = fg$. Express f and g in the standard form. Then if f and g were not both in the form of (3), their product, N, would have a representation distinct from Eq. (1). This contradiction proves the corollary.

Now we are able to complete the proof of Theorem 2.

PROOF OF THEOREM 2. If $2^p - 1 = p_2$ is a prime, the only positive divisors of

$$N = 2^{p-1} p_2$$

are those listed on page 4. Therefore N is the sum of *all* its positive divisors, other than itself, and N is perfect.

The logical structure of the theorems discussed so far is shown in the following diagram. The important theorems are those at the bottom.

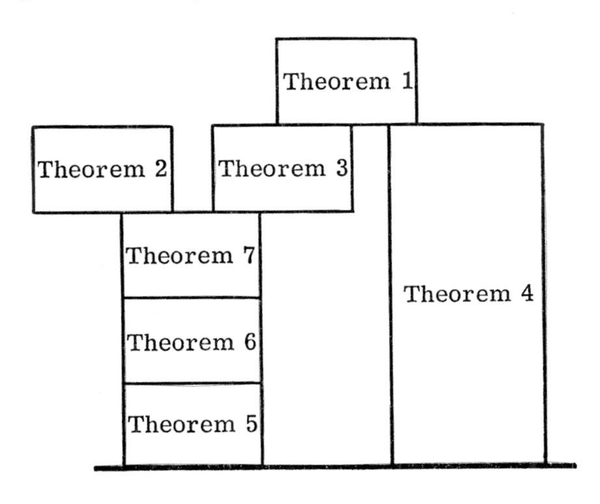

They support the theorems which rest upon them. In general, the important theorems will have many consequences, while Theorem 1, for instance, has almost no consequence of significance.

The proofs of Theorems 3 and 5 will now be given.

3. Euler's Converse Proved

Proof of Theorem 3 (by L. E. Dickson). Let N be an even perfect number given by

$$N = 2^{n-1}F$$

where F is an odd number. Let Σ be the sum of the positive divisors of F. The positive divisors of N include all these odd divisors and their doubles, their multiples of 4, $\cdots$, their multiples of 2^{n-1}. There are no other positive divisors by the corollary of Theorem 7. Since N is perfect we have

$$N = 2^{n-1}F = (1 + 2 + \cdots + 2^{n-1})\Sigma - N$$

or

$$2N = 2^nF = (2^n - 1)\Sigma.$$

Therefore

$$\Sigma = F + F/(2^n - 1), \tag{4}$$

and since Σ and F are integers, so must $F/(2^n - 1)$ be an integer. Thus

$$(2^n - 1)|F$$

and $F/(2^n - 1)$ must be one of the divisors of F. Since Σ is the sum of *all* the positive divisors of F, we see, from Eq. (4), that there can only be two, namely F itself and $F/(2^n - 1)$. But 1 is certainly a divisor of F. Therefore $F/(2^n - 1)$ must equal 1, F must equal $2^n - 1$, and $2^n - 1$ has no other positive divisors. That is, $2^n - 1$ is a prime.

4. Euclid's Algorithm

Proof of Theorem 5 (Euclid's Algorithm). To compute the greatest common divisor of two *positive* integers a and b, Euclid proceeds as follows. Without loss of generality, let $a \leqq b$ and divide b by a:

$$b = q_0a + a_1$$

with a positive quotient q_0, and a remainder a_1 where $0 \leqq a_1 < a$. If $a_1 \neq 0$, divide a by a_1 and continue the process until some remainder, a_{n+1}, equals 0.

$$a = q_1 a_1 + a_2$$
$$a_1 = q_2 a_2 + a_3$$
$$\vdots$$
$$a_{n-2} = q_{n-1} a_{n-1} + a_n$$
$$a_{n-1} = q_n a_n .$$

This must occur, since $a > a_1 > a_2 \cdots > 0$. Then the greatest common divisor, $g = (a, b)$, is given by

$$g = a_n . \tag{5}$$

For, from the first equation, since $g|a$ and $g|b$, we have $g|a_1$. Then, from the second, since $g|a$ and $g|a_1$, we have $g|a_2$. By induction, $g|a_n$, and therefore

$$g \leqq a_n . \tag{6}$$

But, conversely, since $a_n|a_{n-1}$ by the last equation, by working backwards through the equations we find that $a_n|a_{n-2}$, $a_n|a_{n-3}$, $\cdots$, $a_n|a$ and $a_n|b$. Thus a_n is a common divisor of a and b and

$$a_n \leqq g \quad \text{(the greatest)}.$$

With Eq. (6) we therefore obtain Eq. (5). Now, from the next-to-last equation, a_n is a linear combination, with integer coefficients, of a_{n-1} and a_{n-2}. Again working backwards we see that a_n is a linear combination of a_{n-i} and a_{n-i-1} for every i. Finally

$$g = a_n = ma + nb \tag{7}$$

for some integers m and n. If, in Theorem 5, a and b are not both positive, one may work with their absolute values. This completes the proof of Theorem 5, and therefore also the proofs of Theorems 6, 7, 2, 3, and 1.

EXAMPLE: Let $g = (143, 221)$.

Then

$$221 = 1 \cdot 143 + 78,$$
$$143 = 1 \cdot 78 + 65,$$
$$78 = 1 \cdot 65 + 13,$$
$$65 = 5 \cdot 13,$$

and $g = 13$. Now

$$\begin{aligned} 13 &= 78 - 1 \cdot 65 \\ &= 2 \cdot 78 - 1 \cdot 143 \\ &= 2 \cdot 221 - 3 \cdot 143. \end{aligned}$$

The reader will note that in the foregoing proof we have tacitly assumed several elementary properties of the integers which we have not stated explicitly—for example, that $a|b$ and $a|c$ implies $a|b + c$; that $a > 0$, and $b|a$ implies $b \leqq a$, and that the a_1 in $b = q_0a + a_1$ exists and is unique. This latter is called the *Division Algorithm*. For a statement concerning these fundamentals see the Statement on page 217.

It should be made clear that the m and n in Eq. (7) are by no means unique. In fact, for every k we also have

$$\left(m + k\frac{b}{g}\right)a + \left(n - k\frac{a}{g}\right)b = g.$$

Theorem 5 is so fundamental (really more so than that which bears the name, Theorem 7), that it will be useful to list here a number of comments. Most of these are not immediately pertinent to our present problem—that of perfect numbers—and the reader may wish to skip to Sect. 5.

(a) The number $g = (a, b)$ is not only a maximum in the *additive* sense, that is, $d \leqq g$ for every common divisor d, but it is also a maximum in the *multiplicative* sense in that for every d

$$d|g. \tag{8}$$

This is clear, since $d|a$ and $d|b$ implies $d|g$ by Eq. (7).

(b) The number g is also a *minimum* in both additive and multiplicative senses. For if

$$m_1a + n_1b = h \tag{9}$$

for *any* m_1 and n_1, we have, by the same argument,

$$g|h. \tag{10}$$

Then it is also clear that

$$g \leqq \text{every positive } h. \tag{11}$$

(c) This minimum property, (11), may be made the basis of an alternative proof of Theorem 5, one which does not use Euclid's Algorithm. The most significant difference between that proof and the given one is that this alternative proof, at least as usually given, is *nonconstructive*, while Euclid's proof is *constructive*. By this we mean that Euclid actually constructs values of m and n which satisfy Eq. (7), while the alternative proves their *existence*, by showing that their nonexistence would lead to a contradiction. We will find other instances, as we proceed, of analogous situations—both constructive and nonconstructive proofs of leading theorems.

Which type is preferable? That is somewhat a matter of taste. Landau, it is clear from his books, prefers the nonconstructive. This type of proof is often shorter, more "elegant." The constructive proof, on the other hand, is "practical"—that is, it *gives* solutions. It is also "richer," that is, it develops more than is (immediately) needed. The mathematician who prefers the nonconstructive will give another name to this richness—he will say (rightly) that it is "irrelevant."

Which type of proof has the greatest "clarity"? That depends on the *algorithm* devised for the constructive proof. A compact algorithm will often cast light on the subject. But a cumbersome one may obscure it.

In the present instance it must be stated that Euclid's Algorithm is remarkably simple and efficient. Is it not amazing that we find the greatest common divisor of a and b without factoring either number?

As to the "richness" of Euclid's Algorithm, we will give many instances below, (e), (f), (g), and Theorem 10.

Finally it should be noted that some mathematicians regard nonconstructive proofs as objectionable on logical grounds.

(d) Another point of logical interest is this. Theorem 7 is primarily multiplicative in *statement*. In fact, if we delete the "standard form," $p_1 < p_2 < \cdots$, which we can do with no real loss, it appears to be purely multiplicative (in *statement*). Yet the proof, using Theorem 5, involves *addition*, also, since Theorem 5 involves addition. There are alternative proofs of Theorem 7, not utilizing Theorem 5, but, without exception, addition intrudes in each proof somewhere. Why is this? Is it because the demonstration of even one representation in the form of Eq. (1) requires the notion of the *smallest* divisor?

When we come later to the topic of *primitive roots*, we will find another instance of an (almost) purely multiplicative theorem where addition intrudes in the proof.

(e) Without any modification, Euclid's Algorithm may also be used to find $g(x)$, the polynomial of greatest degree, which divides two polynomials, $a(x)$ and $b(x)$. In particular, if $a(x)$ is the derivative of $b(x)$, $g(x)$ will contain all *multiple* roots of $b(x)$. Thus if

$$b(x) = x^3 - 5x^2 + 7x - 3,$$

and

$$b'(x) = a(x) = 3x^2 - 10x + 7,$$

then

$$g(x) = a_1(x) = -\tfrac{8}{9}(x - 1).$$

Therefore

$$(x - 1)^2 | b(x).$$

(f) Without elaboration at this time we note that the quotients, q_i, in the Algorithm may be used to expand the fraction a/b into a *continued fraction*.

Thus

$$\frac{a}{b} = \cfrac{1}{q_0 + \cfrac{1}{q_1 + \cfrac{1}{q_2 + \cdots \cfrac{1}{q_n}}}}, \tag{12}$$

and, specifically,

$$\frac{143}{221} = \cfrac{1}{1 + \cfrac{1}{1 + \cfrac{1}{1 + \cfrac{1}{5}}}}.$$

Similarly from (e) above, we have

$$\frac{3x^2 - 10x + 7}{x^3 - 5x^2 + 7x - 3} = \cfrac{9}{(3x - 5) - \cfrac{8}{(3x - 7)}}.$$

(g) Finally we wish to note that, conversely to Theorem 5, if

$$ma + nb = 1$$

then a is prime to b. But likewise m is prime to n and a and b play the role of the coefficients in *their* linear combination. This reciprocal relationship between m and a, and between n and b, is the foundation of the so called *modulo multiplication groups* which we will discuss later.

Now it is high time that we return to perfect numbers.

5. Cataldi and Others

The first four perfect numbers are

$$2(2^2 - 1),$$
$$2^2(2^3 - 1),$$
$$2^4(2^5 - 1),$$
$$2^6(2^7 - 1).$$

We raise again Conjecture 1. Are there infinitely many perfect numbers? We know of no odd perfect number. Although we have not given him a great deal of background so far, the reader may care to try his hand at:

Exercise 1. If any odd perfect number exists it must be of the form

$$D = (p)^{4a+1}N^2$$

where p is a prime of the form $4m + 1$, $a \geqq 0$, and N is some odd number not divisible by p. In particular, then, D cannot be of the form $4m + 3$. (Descartes, Euler).

Any even perfect number is of the form

$$2^{p-1}(2^p - 1)$$

with p a prime. If there were only a finite number of primes, then, of course, there would only be a finite number of even perfects. Euclid's last contribution is

Theorem 8 (Euclid). *There are infinitely many primes.*

PROOF. If $p_1, p_2, \cdots, p_n$ are n primes (not necessarily consecutive), then since

$$N = p_1 p_2 \cdots p_n + 1$$

is divisible by none of these primes, any prime p_{n+1} which does divide N, (and there must be such by Theorem 7), is a prime not equal to any of the others. Thus the set of primes is not finite.

EXERCISE 2. (A variation on Theorem 8 due to T. J. Stieltjes.) Let A be the product of *any* r of the n primes in Theorem 8, with $1 \leqq r \leqq n$, and let $B = p_1 p_2 \cdots p_n / A$.

Then $A + B$ is prime to each of the n primes.

EXAMPLE: $p_1 = 2$, $p_2 = 3$, $p_3 = 5$. Then

$$2 \cdot 3 \cdot 5 + 1, \qquad 2 \cdot 3 + 5, \qquad 2 \cdot 5 + 3, \qquad 3 \cdot 5 + 2$$

are all prime to 2, 3, and 5.

EXERCISE 3. Let $A_1 = 2$ and A_n be defined recursively by

$$A_{n+1} = A_n^2 - A_n + 1.$$

Show that each A_i is prime to every other A_j. HINT: Show that

$$A_{n+1} = A_1 A_2 \cdots A_n + 1$$

and that what is really involved in Theorem 8 is not so much that the p's are primes, as that they are prime to each other.

EXERCISE 4. Similarly, show that all of the *Fermat Numbers*,

$$F_m = 2^{2^m} + 1$$

for $m = 0, 1, 2, \cdots$, are prime to each other, since

$$F_{m+1} = F_0 F_1 \cdots F_m + 2.$$

Here, and throughout this book, 2^{2^m} means $2^{(2^m)}$, not $(2^2)^m = 2^{2m} = 4^m$.

EXERCISE 5. Show that either the A_i of Exercise 3, or the F_i of Exercise 4, may be used to give an alternative proof of Theorem 8.

Thus there are infinitely many values of $2^p - 1$ with p a prime. If, as Leibnitz erroneously believed, the converse of Theorem 4 were true, that is, if p's primality implied $2^p - 1$'s primality, then Conjecture 1 would follow immediately from Euclid's Theorem 2 and Theorem 8. But the converse of Theorem 4 is false, since already

$$23|2^{11} - 1,$$

a fact given above in disguised form (example of Definition 4).

Definition 5. Henceforth we will use the abbreviation

$$M_n = 2^n - 1.$$

M_n is called a *Mersenne number* if n is a prime.

Skipping over an unknown computer who found that M_{13} was prime, and that $P_5 = 2^{12}M_{13}$ was therefore perfect, we now come to Cataldi (1588). He showed that M_{17} and M_{19} were also primes. Now $M_{19} = 524{,}287$, and we are faced with a leading question in number theory. Given a large number, say $M_{31} = 2147483647$, is it a prime or not?

To show that N is a prime, one could attempt division by $2, 3, \cdots, N - 1$, and if N is divisible by none of these then, of course, it is prime. But this is clearly wasteful, since if N has no divisor, other than 1, which satisfies

$$d \leqq \sqrt{N}$$

then N must be a prime since, if

$$N = fg,$$

f and g cannot both be $> \sqrt{N}$. Further, if we have a table of primes which includes all primes $\leqq \sqrt{N}$, it clearly suffices to use these primes as trial divisors since the *smallest* divisor (>1) of N is always a prime.

Definition 6. If x is a real number, by

$$[x]$$

we mean the greatest integer $\leqq x$.

Examples:

$$1 = [1.5], 2 = [2], 3 = [3.1417],$$

$$-1 = [-\tfrac{1}{2}], 724 = [\sqrt{M_{19}}].$$

To prove that $M_{19} = 524{,}287$ is a prime, Cataldi constructed the first extensive table of primes—up to 750—and he simply tried division of M_{19} by all the primes $<[\sqrt{M_{19}}] = 724$. There are 128 such primes. This was rather laborious, and since M_n increases so very rapidly, it virtually forces the creation of other methods. To estimate the labor involved in proving some M_p a prime by Cataldi's method, we must know the number of primes $< \sqrt{M_p}$.

Definition 7. Let

$$\pi(n)$$

be the number of primes which satisfy $2 \leqq p \leqq n$.

EXAMPLE:

$$\pi(724) = 128.$$

There is no shortage of primes. A brief table shows the trend.

n	$\pi(n)$	
10	4	
10^2	25	
10^3	168	
10^4	1,229	
10^5	9,592	
10^6	78,498	
10^7	664,579	
10^8	5,761,455	
10^9	50,847,534	(D. H. Lehmer)
10^{10}	455,052,511	(D. H. Lehmer)

This brings us to the prime number theorem.

6. THE PRIME NUMBER THEOREM

In Fermat's time (1640), Cataldi's table of primes was still the largest in print. In Euler's time (1738), there was a table, by Brancker, up to 100,000. In Legendre's time (1798), there was a table, by Felkel, up to 408,000.

The distribution of primes is most irregular. For example (Lehmer), there are no primes between 20,831,323 and 20,831,533, while on the other hand (Kraitchik), 1,000,000,000,061 and 1,000,000,000,063 are both primes. No simple formula for $\pi(n)$ is either known, nor can one be expected. But, "in the large," a definite trend is readily apparent, (see the foregoing table), and on the basis of the tables then existing, Legendre (1798, 1808) conjectured, in effect, the Prime Number Theorem.

Definition 8. If $f(x)$ and $g(x)$ are two functions of the real variable x, we say that $f(x)$ is *asymptotic* to $g(x)$, and write it

$$f(x) \sim g(x),$$

if

$$\lim_{x\to\infty} \frac{f(x)}{g(x)} = 1 .$$

Theorem 9. (*The Prime Number Theorem, conjectured by Legendre, Gauss, Dirichlet, Chebyshev, and Riemann; proven by Hadamard and de la Vallée Poussin in 1896*).

$$\pi(n) \sim \frac{n}{\log n} .$$

No easy proof of Theorem 9 is known. The fact that it took a century to prove is a measure of its difficulty. The theorem is primarily one of analysis. Number theory plays only a small role. That some analysis must enter is clear from Definition 8—a *limit* is involved. The *extent* to which analysis is involved is what is surprising. We shall give a proof in Volume II.

For now we wish to make some clarifications. Definition 8 does *not* mean that $f(x)$ is *approximately* equal to $g(x)$. This has no strict mathematical meaning. The definition in no way indicates anything about the difference

$$f(x) - g(x),$$

merely about the *ratio*

$$f(x)/g(x).$$

Thus

$$n^2 + 1 \sim n^2$$

$$n^2 + 100n \sim n^2$$

and

$$n^2 + n^{1.9} \log n \sim n^2$$

are equally true. Which function, on the left, is the best approximation to n^2 is quite a different problem.

If

$$f(x) \sim g(x)$$

and

$$g(x) \sim h(x)$$

then

$$f(x) \sim h(x).$$

Theorem 9 may therefore take many forms by replacing $n/\log n$ by any function asymptotic to it. Thus

Theorem 9_1.

$$\pi(n) \sim \frac{n}{\log n - 1}.$$

Theorem 9_2.

$$\pi(n) \sim \int_2^n \frac{dx}{\log x}.$$

These three versions are all equally true. Which function on the right is the best approximation?

P. Chebyshev (1848) gave both Theorems 9_1 and 9_2, but proved neither. C. F. Gauss, in a letter to J. F. Encke (1849), said that he discovered Theorem 9_2 at the age of 16—that is, in 1793—and that when Chernac's factor table to 1,020,000 was published in 1811 he was still an enthusiastic prime counter. Glaisher describes this letter thus:

"The appearance of Chernac's *Cribum* in 1811 was, Gauss proceeds, a cause of great joy to him; and, although he had not sufficient patience for a continuous enumeration of the whole million, he often employed unoccupied quarters of an hour in counting here and there a chiliad."

EXERCISE 6. Compute $N/\log N - 1$ (natural logarithm, of course!) for $N = 10^n$, $n = 1, 2, \cdots, 10$, and compare the right and left sides of Theorem 9_1.

7. Two Useful Theorems

Before we consider the work of Fermat, it will be useful to give two theorems. The first is an easy generalization of an argument used in the proof of Theorem 4, page 3. We formalize this argument as

Theorem 4_0. *If $x \neq y$, and $n > 0$, then*

$$x - y | x^n - y^n. \tag{13}$$

In particular, if $y = 1$,

$$x - 1 | x^n - 1, \tag{13a}$$

and, if $y = -y$, and n is odd,

$$x + y | x^n + y^n, \qquad (n \text{ is odd}). \tag{13b}$$

The proof is left to the reader.

Theorem 10. *If a, b, and s are positive integers, we write*

$$s^a - 1 = B_a, \qquad s^b - 1 = B_b.$$

Then if $(a, b) = g$,

$$(B_a, B_b) = B_g, \tag{14}$$

and in particular if a is prime to b, then $s - 1$ *is the greatest common divisor of* $s^a - 1$ *and* $s^b - 1$.

PROOF. In computing $g = (a, b)$ by Euclid's Algorithm, the $(m + 1)$st equation (page 9) is

$$a_{m-1} = q_m a_m + a_{m+1}. \tag{15}$$

It follows that

$$B_{a_{m-1}} = Q_m B_{a_m} + B_{a_{m+1}} \tag{16}$$

for some integer Q_m, for the reader may verify that

$$\begin{aligned} B_{a_{m-1}} &= s^{a_{m-1}} - 1 = B_{a_{m+1}} B_{q_m a_m} + B_{q_m a_m} + B_{a_{m+1}} \\ &= \left[(B_{a_{m+1}} + 1) \frac{B_{q_m a_m}}{B_{a_m}} \right] B_{a_m} + B_{a_{m+1}}. \end{aligned}$$

But $B_{a_m} | B_{q_m a_m}$ by Eq. (13a) with $x = s^{a_m}$, and $n = q_m$, and thus

$$(B_{a_{m+1}} + 1) \frac{B_{q_m a_m}}{B_{a_m}}$$

is an integer. Call it Q_m and this proves Eq. (16).

But were we to compute (B_a, B_b) by Euclid's Algorithm, Eq. (16) would be the $m + 1$st equation and the remainder, $B_{a_{m+1}}$, of Eq. (16) corresponds to the remainder, a_{m+1}, of Eq. (15). Therefore if $(a, b) = g$, $(B_a, B_b) = B_g$.

Corollary. *Every Mersenne number,* $M_p = 2^p - 1$, *is prime to every other Mersenne number.*

The correspondence between Eqs. (15) and (16) has an interesting *arithmetic* interpretation. For simplicity, let $s = 2$ and thus $B_a = M_a = 2^a - 1$. Let

$$b = qa + r \tag{17}$$

and

$$M_b = QM_a + M_r. \tag{18}$$

Now M_x, in *binary*, is a string of x ones, and if the division, Eq. (18), is carried out in binary we divide a string of a 1's into a string of b 1's

$$\begin{array}{r|l} & 100001000 \\ \hline 11111 & 11111111111111 \\ & 1111111111 \\ \cline{2-2} & 111 \end{array} \tag{18a}$$

and obtain a remainder of r 1's. On the other hand, the ancient interpretation of Eq. (17) is that a stick b units long is *measured* by a stick a units long, q times, with a remainder r units long.

```
|11111 11111 111|
|11111|11111|
            |111|
```
(17a)

The quotient Q, of Eq. (18), consists of the q marks (bits) made in measuring M_b by M_a !

$$Q: \boxed{0000100001000}$$

8. Fermat and Others

Now we come to Pierre de Fermat. In the year 1640, France was the leading country of Europe, both politically and culturally. The political leader was Cardinal Richelieu. The leading mathematicians were René Descartes, Gérard Desargues, Fermat, and the young Blaise Pascal. In 1637, Descartes had published *La Geometrie*, and in 1639 the works of Desargues and Pascal on projective geometry had appeared. From 1630 on, Father Marin Mersenne, a diligent correspondent (with an inscrutable handwriting) had been sending challenge problems to Descartes, Fermat, Frenicle, and others concerning perfect numbers and related concepts. By his perseverance, he eventually persuaded all of them to work on perfect numbers.

At this time M_2, M_3, M_5, M_7, M_{13}, M_{17}, and M_{19} were known to be prime. But

$$M_{11} = 23 \cdot 89,$$

and Fermat found that

$$47 | M_{23} .$$

The obvious numerical relationship between $p = 11$ and the factors 23 and 89, in the first instance, and between 23 and 47 in the second, may well have suggested to Fermat the following

Theorem 11 (Fermat, 1640). *If $p > 2$, any prime which divides M_p must be of the form $2kp + 1$ with $k = 1, 2, 3, \cdots$.*

At the same time Fermat found:

Theorem 12 (Fermat, 1640). *Every prime p divides $2^p - 2$:*

$$p | 2^p - 2. \tag{19}$$

These two important theorems are closely related. That Theorem 11 implies Theorem 12 is easily seen. Since the product of two numbers of the form $2kp + 1$ is again of that form, it is clear by induction that Theorem 11 implies that *all* divisors of M_p are of that same form. Therefore M_p itself equals $2Kp + 1$ for some K, and thus $M_p - 1$ is a multiple of p. And this is Theorem 12. The case $p = 2$ is obvious.

But conversely, Theorem 12 implies Theorem 11. For let a prime q divide M_p. Then

$$q | 2^p - 1, \tag{20}$$

and by Theorems 12 and 6,

$$q | 2^{q-1} - 1. \tag{21}$$

Now by Theorem 10, $(2^p - 1, 2^{q-1} - 1) = 2^g - 1$ where $g = (p, q - 1)$. Since $q > 1$, we have from Eqs. (20) and (21) that $g > 1$. But since p is a prime, we therefore have $p | q - 1$, or $q = sp + 1$. Finally if s were odd, q would be even and thus not prime. Therefore q is of the form $2kp + 1$. To prove Theorems 11 and 12, it therefore will suffice to prove one of the two.

Several months after Fermat announced these two theorems (in a letter to Frenicle), he generalized Theorem 12 to the most important

Theorem 13 (Fermat's Theorem). *For every prime p and any integer a,*

$$p | a^p - a. \tag{22}$$

This clearly implies Theorem 12, and is itself equivalent to

Theorem 13_1. *If $p \nmid a$, then*

$$p | a^{p-1} - 1. \tag{23}$$

For if $p | a(a^{p-1} - 1)$ and $p \nmid a$ then by Theorem 6, $p | a^{p-1} - 1$. The converse implication is also clear. Nearly a century later, Euler generalized Theorem 13_1 and in doing so he introduced an important function, $\phi(n)$.

Definition 9. If n is a positive integer, the number of positive integers prime to n and $\leqq n$ is called $\phi(n)$, *Euler's phi function*. There are therefore $\phi(n)$ solutions m of the system:

$$\begin{cases} (m, n) = 1 \\ 1 \leqq m \leqq n. \end{cases}$$

EXAMPLES:

$$\phi(1) = 1, \quad \phi(2) = 1, \quad \phi(3) = 2, \quad \phi(4) = 2, \quad \phi(5) = 4,$$

$$\phi(6) = 2, \quad \phi(7) = 6, \quad \phi(8) = 4, \quad \phi(9) = 6, \quad \phi(10) = 4.$$

For any prime, p, $\phi(p) = p - 1$.

Theorem 14 (Euler). *For any positive integer m, and any integer a prime to m,*

$$m | a^{\phi(m)} - 1. \tag{24}$$

Later we will prove Theorem 14, and since, for a prime p, $\phi(p) = p - 1$, this will also prove the special case Theorem 13_1. That will complete the proofs of Theorems 13, 12, and 11. For the moment let us consider the significance of Fermat's Theorem 11 for the perfect number problem.

The first Mersenne number we have not yet discussed is M_{29}. To determine whether it is a prime, it is *not* necessary to attempt division by 3, 5, 7, etc. The only *possible* divisors are those of the form $58k + 1$. For $k = 1, 2, 3$, and 4 we have $58k + 1 = 59, 117, 175$, and 233. But $59 \nmid M_{29}$. Again, 117 and 175 are not primes and therefore need not be tried, since the *smallest* divisor (>1) must be a prime. Finally $233 | M_{29}$. Thus we find that $M_{29} = 536{,}870{,}911$ is composite with only 2 trial divisions.

Exercise 7. Assume that $p = 16035002279$ is a prime, (which it is), and that $q = 32070004559$ divides M_p, (which it does). Prove that q is a prime.

Exercise 8. Verify that

$$3 \cdot 74 + 1 | M_{37}.$$

(When we get to Gauss's conception of a *residue class*, such computations as that of this exercise will be much abbreviated.)

It has been similarly shown that M_{41}, M_{43}, M_{47}, M_{53}, and M_{59} are also composite. Up to $p = 61$, there are nine Mersenne primes, that is, M_p for $p = 2, 3, 5, 7, 13, 17, 19, 31$, and 61. These nine primes are listed in the table on page 22, together with four other columns.

The first two columns are

$$s_p = [\sqrt{M_p}] \tag{25}$$

and

$$c_p = \pi(s_p). \tag{26}$$

The number c_p is the number of trial divisions—à la Cataldi (see page 14) needed to prove M_p a prime.

Definition 10. By $\pi_{a,b}(n)$ is meant the number of primes of the form $ak + b$ which are $\leqq n$.

Examples:

$\pi_{4,1}(50) = 6$; the six primes being 5, 13, 17, 29, 37, 41
$\pi_{4,3}(50) = 8$; the eight primes being 3, 7, 11, 19, 23, 31, 43, 47
$\pi_{8,1}(10^6) = 19552$
$\pi_{8,3}(10^6) = 19653$

$$\pi_{8,5}(10^6) = 19623$$
$$\pi_{8,7}(10^6) = 19669.$$

By Theorem 11, the only primes which may divide M_p are those counted by the function $\pi_{2p,1}(n)$. The next column of the table is

$$f_p = \pi_{2p,1}(s_p). \tag{27}$$

The last column, e_p, we will explain later. (Mnemonic aid: c_p means "Cataldi," f_p means "Fermat," e_p means "Euler.")

TABLE OF THE FIRST NINE MERSENNE PRIMES

p	M_p	s_p	c_p	f_p	e_p
2	3	1	0	0	0
3	7	2	1	0	0
5	31	5	3	0	0
7	127	11	5	0	0
13	8,191	90	24	2	1
17	131,071	362	72	4	3
19	524,287 (Cataldi, 1588)	724	128	6	3
31	2,147,483,647 (Euler, 1772)	46,340	4,792	157	84
61	2,305,843,009,213,693,951	$1.5 \cdot 10^9$	$75 \cdot 10^6$*	$1.25 \cdot 10^6$**	$0.62 \cdot 10^6$**

* Estimated, using Theorem 9.
** Estimated, using Theorem 16.

We see in the table that had Cataldi known Theorem 11, the 128 divisions which he performed in proving M_{19} a prime could have been reduced to 6; $f_{19} = 6$.

EXERCISE 9. Identify the two primes in f_{13}, namely those of the form $26k + 1$ which are <90. Also identify the 4 primes in f_{17}.

We inquire now whether the ratio f_p/c_p will always be as favorable as the instances cited above. More generally, how does $\pi_{a,b}(n)$ compare with $\pi(n)$? Since $ak + b$ is divisible by $g = (a, b)$ it is clear that the form $ak + b$ cannot contain infinitely many primes *unless* b is prime to a. But suppose $(a, b) = 1$? If we hold a fixed there are $\phi(a)$ values of b which are $<a$ and prime to a. Does each such form possess infinitely many primes? Two famous theorems answer this question:

Theorem 15 (Dirichlet, 1837). *If $(a, b) = 1$, there are infinitely many primes of the form $ak + b$.*

A stronger theorem which implies Theorem 15 (and also Theorem 9) is

Theorem 16 (de la Vallée Poussin, 1896). *If $(a, b) = 1$, then*

$$\pi_{a,b}(n) \sim \frac{1}{\phi(a)} \frac{n}{\log n} \sim \frac{1}{\phi(a)} \pi(n), \tag{28}$$

or, equivalently, for any two numbers prime to a, b′ and b″, we have

$$\pi_{a,b'}(n) \sim \pi_{a,b''}(n). \tag{29}$$

We postpone the proof of Theorem 15 to Volume II, but a special case which we need later is proven in Section 36. The more difficult Theorem 16 will be used as a *guide* in the following investigations but will not be used logically and will not be proven. We note that although Eq. (28) is an *asymptotic* law, we may nonetheless employ it for even modest values of n with a usable accuracy. Thus $\phi(38) = 18$; more generally, for any prime p, $\phi(2p) = p - 1$. Then $\pi(s_{19}) = 128$ and $\frac{1}{18}\pi(s_{19}) = 7.1$. The number sought is $\pi_{38,1}(s_{19}) = f_{19} = 6$, a reasonable agreement considering the smallness of the numbers involved. Generally we should expect

$$f_p \approx \frac{1}{p-1} c_p \tag{30}$$

but it is clear that this is not an exact statement, since we give no bound on the error.

Exercise 10. The ratio s_p/c_p may be regarded as a measure of the improvement introduced by Cataldi by his procedure of using only *primes* as trial divisors (page 14). Similarly, c_p/f_p measures the improvement made by Fermat. Now note that the second ratio runs about 3 times the first, so that we may say that Fermat's improvement was the larger of the two. Interpret this constant (≈ 3) as $2/\log 2$ by using the estimates for c_p and f_p suggested by Theorems 9 and 16. Evaluate this constant to several decimal places.

9. Euler's Generalization Proved

We now return to Euler's Theorem 14,

$$m | a^{\phi(m)} - 1, \qquad (a, m) = 1$$

which we will prove by the use of the important

Theorem 17. *Let $m > 1$. Let a_i, $1 \leqq i \leqq \phi(m)$, be the $\phi(m)$ positive integers $< m$ and prime to m. Let a be any integer prime to m. Let the $\phi(m)$ products, $aa_1, aa_2, \cdots, aa_{\phi(m)}$ be divided by m, giving*

$$aa_i = q_i m + r_i \tag{31}$$

with

$$0 \leqq r_i < m.$$

Then the $\phi(m)$ values of r_i are distinct, and are equal to the $\phi(m)$ values of a_i in some rearrangement.

Proof of Theorem 17. Since a and a_i are both prime to m, so is their product—by Theorem 5, Corollary. Therefore, from Eq. (31), r_i is also prime to m and thus is equal to one of the a_i. If $r_i = r_j$ we have from Eq. (31),

$$a(a_i - a_j) = (q_i - q_j)m$$

Thus from Theorem 6, since $(a, m) = 1$,

$$m | a_i - a_j$$

or $a_i = a_j$. Thus the r_i are all distinct.

PROOF OF THEOREM 14 (by Ivory). The product of any two equations in Eq. (31) is

$$a^2 a_i a_j = Qm + r_i r_j$$

for some integer Q, and by induction, the product of all $\phi(m)$ equations in Eq. (31) can be written

$$a^{\phi(m)} a_1 a_2 \cdots a_{\phi(m)} - r_1 r_2 \cdots r_{\phi(m)} = Lm$$

for some integer L. But (Theorem 17) the product of all the r_i equals the product of all the a_i. Since

$$(a^{\phi(m)} - 1) a_1 a_2 \cdots a_{\phi(m)}$$

is divisible by m, and each a_i is prime to m, by Theorem 6

$$m | a^{\phi(m)} - 1.$$

This completes the proofs of Theorems 14, 13_1, 13, 12, and 11.

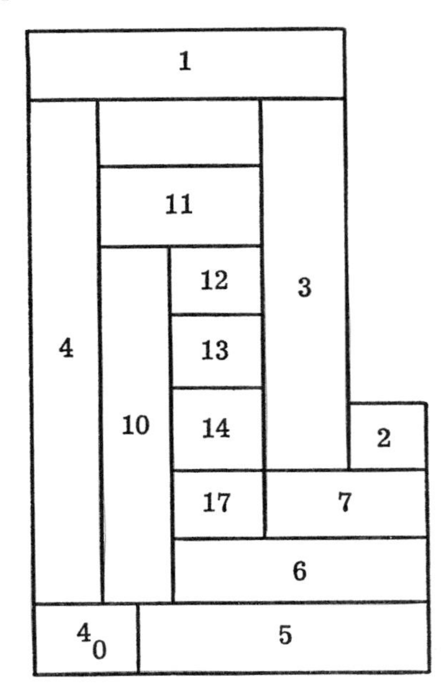

Our logical structure so far (not including Theorem 8 and the unproven Theorems 9, 15 and 16) is given by the diagram on the previous page.

10. Perfect Numbers, II

In the previous sections we have attempted to look at the perfect numbers thru the eyes of Euclid, Cataldi and Fermat, and to examine the consequences of these several inspections. In the next section we take up other important implications which were discovered by Euler. The reader may be inclined to think that we have no sincere interest in the perfect numbers, as such, but are merely using them as a vehicle to take us into the fundamentals of number theory. We grant a grain of truth to this allegation—but only a grain. For consider the following:

If N is perfect it equals the sum of its divisors other than itself.

$$N = 1 + d_1 + d_2 + \cdots + d_n$$

Dividing by N, we find that the sum of the *reciprocals* of the divisors, other than 1, is equal to 1.

$$1 = \frac{1}{N} + \frac{1}{d_1} + \frac{1}{d_2} + \cdots + \frac{1}{d_n}.$$

For $P_2 = 28$, we have, for instance,

$$1 = \frac{1}{7} + \frac{1}{14} + \frac{1}{28} + \frac{1}{4} + \frac{1}{2}.$$

Now write these fractions in binary notation. Since 7 (decimal) = 111 (binary), we have

$$\begin{aligned} \tfrac{1}{7} &= .001001001001\cdots \\ \tfrac{1}{14} &= .000100100100\cdots \quad &&\text{(shift right one place)} \\ \tfrac{1}{28} &= .000010010010\cdots \quad &&\text{(shift right one place)} \\ \tfrac{1}{4} &= .010000000000\cdots \\ \tfrac{1}{2} &= .100000000000\cdots \\ \text{sum} = 1 &= .111111111111\cdots \end{aligned}$$

The fractions not only add to 1, but do so without a single carry! And as it is with 28, so is it with 496. Is that not *perfection*—of a sort?

11. Euler and M_{31}

We continue to examine the Mersenne numbers, M_p, and our attempt to determine which of these numbers are prime. In Theorem 11 we found that any prime divisor of M_p is *necessarily* of the form $2kp + 1$. We now seek a sufficient condition—that is, given a prime p and a second prime $q = 2kp + 1$, what criterion will suffice to guarantee that $q|M_p$? Consider the first case, $k = 1$. Given a prime p, $q = 2p + 1$ may be a prime, as for

$p = 3$, or it may not, as for $p = 7$. If it is, q may divide M_p, as

$$23|M_{11} \text{ and } 47|M_{23}$$

or it may not, as

$$11\nmid M_5 \text{ and } 59\nmid M_{29}{}^{*}.$$

What distinguishes these two classes of q? To help us discover the criterion, consider a few more cases:

$$7|M_3{}^{**} \quad \text{and} \quad 167|M_{83}$$

but

$$83\nmid M_{41} \quad \text{and} \quad 107\nmid M_{53}.$$

The reader may verify (in all *these* cases) that if p is of the form $4m + 3$ and thus $q = 8m + 7$, then $q|M_p$, whereas if p is of the form $4m + 1$ and thus $q = 8m + 3$, then $q\nmid M_p$. Does this rule always hold?

Consider the question in a more general form. Let $q = 2Q + 1$ be a prime with Q *not necessarily* a prime. When does

$$q|2^Q - 1?$$

By Fermat's Theorem 13_1 we had

$$q|2^{2Q} - 1,$$

and factoring the right side:

$$q|(2^Q - 1)(2^Q + 1),$$

we find from Theorem 6, Corollary that either

$$q|2^Q - 1$$

or

$$q|2^Q + 1.$$

It cannot divide them both since their difference is only 2. Which does it divide? To give the answer in a compact form we write the class of integers $8k + 7$ as $8k - 1$ and the class $8k + 5$ as $8k - 3$. Then we have

Theorem 18. *If* $q = 2Q + 1$ *is prime, then*

$$q|2^Q - 1 \quad \text{if} \quad q = 8k \pm 1, \tag{32}$$

and

$$q|2^Q + 1 \quad \text{if} \quad q = 8k \pm 3. \tag{33}$$

---

* Nonetheless M_{29} is composite, since $233|M_{29}$.

** Nonetheless M_3 is prime, since $7 = M_3$.

In view of the discussion above we can at once write the

Corollary. *If $p = 4m + 3$ is a prime, with $m > 0$, and if $q = 2p + 1$ is also a prime, then $q|M_p$—and thus $2^{p-1}M_p$ is not perfect.*

Like Fermat's Theorem 12, we will not prove Theorem 18 directly, but deduce it from a more general theorem. This time, however, the generalization is by no means as simple, and we shall not prove Theorem 18 until Section 17. For now we deduce a second important consequence.

Theorem 19. *Every divisor of M_p, for $p > 2$, is of the form $8k \pm 1$.*

PROOF. Let $q = 2Q + 1$ be a prime divisor of M_p. Then

$$q|2M_p = 2^{p+1} - 2 = N^2 - 2 \tag{34}$$

where

$$N = 2^{(p+1)/2}.$$

Thus

$$2 = N^2 - Kq$$

for some integer K. Then

$$2^2 = N^4 - K_2q$$

for some integer K_2, and, by induction

$$2^Q = N^{2Q} - Lq.$$

Now $q\nmid N$, since $q\nmid 2$, and thus, by Fermat's Theorem, $q|N^{2Q} - 1$. Therefore $q|2^Q - 1$, and, by Theorem 18, q must be of the form $8k \pm 1$. Finally, since the product of numbers of the form $8k \pm 1$ is again of that form, *all* divisors of M_p are of the form $8k \pm 1$.

We were seeking a sufficient condition for $q|M_p$ and found one in the corollary of the previous theorem. Here instead we have another necessary condition. Let us return to the table on page 22. We may now define e_p, the last column. From the primes counted by $f_p = \pi_{2p,1}(s_p)$, we delete those of the form $8k \pm 3$. By Theorem 19 only the remaining primes can qualify to be the smallest prime divisor of M_p. We call the number of these primes e_p.

As an example, consider M_{31}. For nearly 200 years, Cataldi's M_{19} had been the largest known Mersenne prime. To test M_{31}, we examine the primes which are $<46{,}340$, of the form $62k + 1$, and of the form $8k \pm 1$. Let $k = 4j + m$ with $m = 0, 1, 2,$ and 3. Then the primes of the form $62k + 1$ are of four types:

$$\begin{aligned}
248j + 1 &= 8(31j) + 1 \\
248j + 63 &= 8(31j + 8) - 1 \\
248j + 125 &= 8(31j + 16) - 3 \\
248j + 187 &= 8(31j + 23) + 3.
\end{aligned}$$

The last two types we eliminate, leaving

$$e_{31} = \pi_{248,\,1}(s_{31}) + \pi_{248,\,63}(s_{31}).$$

Euler found that no prime q satisfied

$$q < 46{,}340^*$$

$$q = \begin{cases} 248k + 1 & \text{or} \\ 248k + 63 \end{cases}$$

and

$$q \mid M_{31}.$$

Thus $M_{31} = 2147483647$ was the new largest known prime. It remained so for over 100 years.

Exercise 11. Show that if $p = 4m + 3$, $q = 2kp + 1$, and $q \mid M_p$, then $k = 4r$ or $k = 4r + 1$. If $p = 4m + 1$, and $q \mid M_p$, then $k = 4r$ or $k = 4r + 3$.

Exercise 12. Show that $4p + 1$ never divides M_p.

Exercise 13. Show that if $p = 4m + 3$,

$$e_p = \pi_{8p,\,1}(s_p) + \pi_{8p,\,2p+1}(s_p),$$

while if $p = 4m + 1$,

$$e_p = \pi_{8p,\,1}(s_p) + \pi_{8p,\,6p+1}(s_p).$$

Exercise 14. Show that e_p is "approximately" one half of f_p. Compare the actual values of c_{31}, f_{31}, and e_{31} on page 22 with estimates obtained by Theorems 9_1 and 16.

Exercise 15. Identify the 3 primes in e_{19}.

A glance at M_{61}, the last line of the table on page 22, shows that a radically different technique is needed to go much further. Euler's new necessary condition, e_p, only helps a little. But the *theory* underlying e_p is fundamental, as we shall see.

The other advance of Euler, Theorem 18, Corollary, seems of more (immediate) significance for the perfect number problem. It enables us to identify many M_p as composite quite quickly. For the following primes $p = 4m + 3$, $q = 2p + 1$ is also a prime: $p = 11, 23, 83, 131, 179, 191, 239, 251, 359, 419, 431, 443, 491, 659, 683, 719, 743, 911, \cdots$. All these M_p are therefore composite.

In Exercise 12, we saw that $4p + 1 \nmid M_p$. But if $p = 4m + 1$, then $q = 6p + 1 = 8(3m) + 7$ is not excluded by Theorem 19. Again we ask,

* Note that Brancker's table of primes sufficed. It existed then and included primes <100,000—see page 15.

for which primes $p = 4m + 1$ and primes $q = 6p + 1$, does $q|M_p$? But this time the answer is considerably more complicated than was the criterion for $q = 2p + 1$ above. A short table is offered the reader:

$q = 6p + 1 \mid M_p$	$q = 6p + 1 \nmid M_p$
$p = 5, 37, 73, 233$	$p = 13, 17, 61, 101, 137, 173, 181$

Exercise 16. Can you find the criterion which distinguishes these two classes of q? This was probably first found (at least in effect) by F. G. Eisenstein. It is usually stated that the three greatest mathematicians were Archimedes, Newton and Gauss. But Gauss said the three greatest were Archimedes, Newton and Eisenstein! The criterion is given on page 169.

12. Many Conjectures and Their Interrelations

So far we have given only one conjecture. But recall the definitions of conjecture and open question given on page 2. Since by Open Question 1 we indicate a lack of serious evidence for the existence of odd perfects, it is clear that if we nonetheless conjecture that there are infinitely many perfects, what we really have in mind is the stronger

Conjecture 2. *There are infinitely many Mersenne primes.*

Contrast this with

Conjecture 3. *There are infinitely many Mersenne composites, that is, composites of the form $2^p - 1$, with p a prime.*

Is this a conjecture? Yes, it is. It has never been proven. It is clear that *at least* one of these two conjectures must be true.

By Theorem 18, Corollary, Conjecture 3 would follow from the stronger

Conjecture 4. *There are infinitely many primes $p = 4m + 3$ such that $q = 2p + 1$ is also prime.*

But this is also unproven—although here we may add that the evidence for this conjecture is quite good. We listed on page 28 some small p of this type. Much larger p's of this type are also known. Some of these are $p = 16035002279, 16045032383, 16048973639, 16052557019, 16086619079, 16118921699, 16148021759, 16152694583, 16188302111$, etc.

For any of these p, $q = 2p + 1|M_p$, and M_p is a number, which if written out in decimal, would be nearly five billion digits long. Each such number would more than fill the telephone books of all five boroughs of New York City. Imagine then, if Cataldi were alive today, and if he set himself the task of proving these M_p composite—by *his* methods! Can't you see the picture—the ONR contract—the thousands of graduate assistants gainfully employed—the Beneficial Suggestion Committee, etc.? But we are digressing.

Conjecture 4 also implies the weaker

Conjecture 5. *There are infinitely many primes p such that* $q = 2p + 1$ *is also prime. Or, equivalently, there are infinitely many integers n such that* $n + 1$ *is prime, and n is twice a prime.*

Conjecture 5 is very closely related* to the famous

Conjecture 6 (Twin Primes). *There are infinitely many integers n such that* $n - 1$ *and* $n + 1$ *are both primes.*

While more than one hundred thousand of such twins are known, e.g., $n = 4, 6, 12, 18, 30, \cdots, 1000000000062, 1000000000332, \cdots,$ $140737488353508, 140737488353700$, a proof of the conjecture is still awaited. Yet it is probable that a much stronger conjecture is true, namely

Conjecture 7 (Strong Conjecture for Twin Primes). *Let* $z(N)$ *be the number of pairs of twin primes,* $n - 1$ *and* $n + 1$, *for* $5 \leqq n + 1 \leqq N$. *Then*

$$z(N) \sim 1.3203236 \int_2^N \frac{dn}{(\log n)^2} \cdot \tag{35}$$

The constant in Eq. (35) is not empirical but is given by the infinite product

$$1.3203236 \cdots = 2 \prod_{p=3}^{\infty} \left\{1 - \frac{1}{(p-1)^2}\right\} \tag{35a}$$

taken over all odd primes.

In Exercise 37S, page 214, we will return to this conjecture. It is known to be intimately related to the famous

Conjecture 8 (Goldbach Conjecture). *Every even number* >2 *is the sum of two primes.*

EXAMPLES:

$$\begin{aligned} 4 &= 2 + 2 \\ 6 &= 3 + 3 \\ 8 &= 3 + 5 \\ 10 &= 5 + 5 = 3 + 7, \text{ etc.} \end{aligned}$$

Returning to Conjecture 5, we will indicate now that it is also related

* By "related" we mean here that the heuristic arguments for the two conjectures are so similar that if we succeed in proving one conjecture, the other will almost surely yield to the same technique.

to Artin's Conjecture and to Fermat's Last Theorem, but it would be too digressive to give explanations at this point.

We had occasion, in the proof of Theorem 19, to use the fact that

$$2M_p = N^2 - 2$$

for some N. Thus Conjecture 2 implies the much weaker

Conjecture 9. *There are infinitely many n for which $n^2 - 2$ is twice a prime.*

This is clearly related to

Conjecture 10. *There are infinitely many primes of the form $n^2 - 2$.*

While more than 15,000 of such primes are known, e.g. $n = 2, 3, 5, 7, 9, \cdots, 179965, \cdots$, a proof of the conjecture is still awaited. Yet it is probable that a much stronger conjecture is true, namely

Conjecture 11. *Let $P_{-2}(N)$ be the number of primes of the form $n^2 - 2$ for $2 \leqq n \leqq N$. Then*

$$P_{-2}(N) \sim 0.9250272 \int_2^N \frac{dn}{\log n}. \tag{36}$$

On page 48 we will return to this conjecture. It is known to be related to

Conjecture 12. *Let $P_1(N)$ be the number of primes of the form $n^2 + 1$ for $1 \leqq n \leqq N$. Then*

$$P_1(N) \sim 0.6864067 \int_2^N \frac{dn}{\log n}. \tag{37}$$

As in Eq. (35), the constants in Eqs. (36) and (37) are given by certain infinite products. But we must postpone their definition until we define the *Legendre Symbol*.

Exercise 17. On page 29 there are several large primes p for which $q = 2p + 1$ is also prime. These were listed to illustrate Conjecture 4. Now show that the q's also illustrate Conjecture 10.

But we do not want to leave the reader with the impression that number theory consists primarily of unsolved problems. If Theorems 18 and 19 have unleashed a flood of such problems for us, they will also lead to some beautiful theory. To that we now turn.

13. Splitting the Primes into Equinumerous Classes

Definition 11. Let A and B be two classes of positive integers. Let $A(n)$ be the number of integers in A which are $\leqq n$; and let $B(n)$ be similarly defined. If

$$A(n) \sim B(n)$$

we say A and B are *equinumerous.*

By this definition and Theorem 16 the four classes of primes: $8k. + 1$, $8k - 1$, $8k + 3$, and $8k - 3$ are all equinumerous. Now Theorem 18 stated that primes $q = 2Q + 1$ divide $2^Q - 1$ if they are of the form $8k + 1$ or $8k - 1$. Otherwise they divide $2^Q + 1$. Therefore the two classes of primes which satisfy

$$q|2^Q - 1 \quad \text{and} \quad q|2^Q + 1$$

are also equinumerous.

We expressed the intent (page 27) to prove Theorem 18 not directly, but, following the precedent:

$$\text{Theorem 13} \rightarrow \text{Theorem 12},$$

to deduce it from the general case. The difficulty is that the generalization is not at all obvious. For the base 3, there is

Theorem 20. *If* $q = 2Q + 1 \neq 3$ *is a prime, then*

$$q|3^Q - 1 \quad \textit{if} \quad q = 12k \pm 1, \tag{38}$$

and

$$q|3^Q + 1 \quad \textit{if} \quad q = 12k \pm 5. \tag{39}$$

Here, again, we find the primes, (not counting 2 and 3), split into equinumerous classes. But this time the split is along quite a different cleavage plane—if we may use such crystallographic language. Thus $7|2^3 - 1$, while $7|3^3 + 1$.

Since primes of the form $8k + 1$ are either of the form $24k + 1$ or of the form $24k + 17$; and since primes of the form $12k - 5$ are either of the form $24k + 7$ or of the form $24k + 19$; etc., the reader may verify that Theorems 18 and 20 may be combined into the following diagram:

For $q = 24k + b = 2Q + 1 =$ prime:

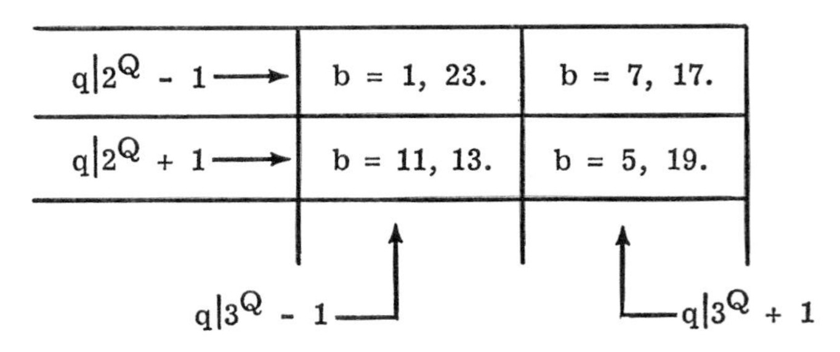

There are, of course, 8 different b's, since $\phi(24) = 8$. It will be useful for the reader at this point, to know a formula of Euler for his phi function. In Sect. 27, when we give the phi function more systematic treatment, we will prove this formula. If N is written in the standard form, Eq. (1), then

$$\phi(N) = N\left(1 - \frac{1}{p_1}\right)\left(1 - \frac{1}{p_2}\right)\cdots\left(1 - \frac{1}{p_n}\right).$$

As an example

$$\phi(24) = 24(1 - \tfrac{1}{2})(1 - \tfrac{1}{3}) = 8.$$

But this does not end the problem of the generalization. Still another base, e.g., 5, 6, 7, etc., will introduce still another cleavage plane. The problem is this: What criterion determines which of the odd primes q, (which do not divide a), divide $a^Q - 1$, and which of them divide $a^Q + 1$? By Theorem 13_1 exactly one of these conditions must exist.

14. Euler's Criterion Formulated

The change of the base from 2 to 3 *changes* the divisibility laws from Eqs. (32) and (33) in Theorem 18 to Eqs. (38) and (39) in Theorem 20. Euler discovered what remains *invariant*. In the proof of Theorem 19 the following implication was used: If there is an N such that $q|N^2 - 2$, then $q|2^Q - 1$. The reader may verify that the number 2 plays no critical role in this argument, so that we can also say that if there is an N such that $q|N^2 - a$, and if $q\nmid a$, then $q|a^Q - 1$. The implication comes from Fermat's Theorem 13_1, and the invariance stems from the invariance in that theorem.

Now Euler found that the converse implication is also true. Thus we will have

Theorem 21 (Euler's Criterion). *Let a be any integer, (positive or negative), and let $q = 2Q + 1$ be a prime which does not divide a. If there is an integer N such that*

$$q|N^2 - a, \quad \textit{then} \quad q|a^Q - 1.$$

If there is no such N, then $q|a^Q + 1$. It follows that the converses of the last two sentences are also true.

Before we prove this theorem, it will be convenient to rewrite it with a "notational change" introduced by Legendre.

Definition 12 (Legendre Symbol—the current, but not the original definition). If q is an odd prime, and a is any integer, then the Legendre Symbol $\left(\frac{a}{q}\right)$ has one of three values. If $q|a$, then $\left(\frac{a}{q}\right) = 0$. If not, then $\left(\frac{a}{q}\right) = +1$ if there is an N such that $q|N^2 - a$, and $\left(\frac{a}{q}\right) = -1$ if there is not.

EXAMPLES:

$$\left(\frac{2}{7}\right) = +1 \qquad \text{since } 7 \mid 3^2 - 2.$$

$$\left(\frac{2}{5}\right) = -1.$$

$$\left(\frac{1}{q}\right) = +1 \qquad \text{since, for every } q,\ q \mid 1^2 - 1.$$

$$\left(\frac{a^2}{q}\right) = +1 \qquad \text{if } q \nmid a, \text{ since, for every } q,\ q \mid a^2 - a^2.$$

Now we may rewrite Euler's Criterion as

Theorem 21_1 . *If $q = 2Q + 1$ is a prime, and a is any integer,*

$$q \,\Big|\, a^Q - \left(\frac{a}{q}\right). \tag{40}$$

We may remark that usually Euler's Criterion is presented as a method of evaluating $\left(\frac{a}{q}\right)$ by determining whether $q \mid a^Q - 1$ or not. The reader may note that we are approaching Euler's Criterion from the opposite direction. The fact is, of course, that Euler's Criterion is a two-way implication, and may be used in either direction.

EXERCISE 18. From Theorems 18 and 21_1 show that for all odd primes p,

$$\left(\frac{2}{p}\right) = (-1)^{(p^2-1)/8}. \tag{41}$$

Likewise

$$\left(\frac{2}{p}\right) = (-1)^{\left[\frac{p+1}{4}\right]} \tag{42}$$

where the square bracket, [ ], is as defined in Definition 6.

EXERCISE 19. Determine *empirically* the "cleavage plane" for $q \mid 5^Q \pm 1$, which is mentioned on page 33, by determining empirically the classes of primes q which divide $N^2 - 5$, and those which do not. That is, factor $N^2 - 5$ for a moderate range of N, and conjecture the classes into which the prime divisors fall. You will be able to *prove* your conjecture after you learn the *Quadratic Reciprocity Law.*

EXERCISE 20. On the basis of your answer to the previous exercise, extend the diagram on page 32 to three dimensions, with the three cleavage planes, $2^Q \pm 1$, $3^Q \pm 1$, and $5^Q \pm 1$. In each of the eight cubes there will be four values of b, corresponding to four classes of primes, $q = 120k + b$. All together there will be 32 classes, corresponding to $\phi(120) = 32$.

15. Euler's Criterion Proved

Our proof of Theorem 21_1 will be based upon a theorem related to Theorem 17.

Theorem 22. *Let q be prime, and let a_i, $i = 1, 2, \cdots, q - 1$, be the positive integers $< q$. Let a be any integer prime to q. Given any one of the a_i, there is a* unique *j such that*

$$q \mid a_i a_j - a. \tag{43}$$

Proof. By Euclid's Eq. (7), page 9, there is an m and an n such that

$$ma_i + nq = 1,$$

or

$$maa_i + naq = a. \tag{44}$$

Since $(m, q) = 1$, we have $q \nmid ma$ and if we divide ma by q we obtain

$$ma = sq + a_j \tag{45}$$

for some j and some s. From Eqs. (44) and (45),

$$q \mid a_i a_j - a.$$

Now, for any k such that

$$q \mid a_i a_k - a,$$

we have

$$q \mid a_i(a_k - a_j),$$

and, since $q \nmid a_i$, we have $q \mid (a_k - a_j)$, that is, $k = j$.

Now we can prove Theorem 21_1.

Proof of Theorem 21_1 (by Dirichlet). Assume first that $\left(\frac{a}{q}\right) = -1$. With reference to Definition 12, this implies that the j and i in Eq. (43) can never be equal. Therefore, by Theorem 22, the $2Q$ integers a_i must fall into Q pairs, and each pair satisfies an equation:

$$a_i a_j = a + Kq \tag{46}$$

for some integer K. The product of these Q equations is therefore

$$(2Q)! = a^Q + Lq$$

for some integer L. Therefore

$$\left(\frac{a}{q}\right) = -1 \quad \text{implies} \quad q|a^Q - (2Q)!. \tag{47}$$

Now assume $\left(\frac{a}{q}\right) = +1$. Then $q|N^2 - a$ for some N, and, since $q \nmid N$ we may write $N = sq + a_r$ for some s and r. Therefore

$$q|a_r^2 - a. \tag{48}$$

If, for any t,

$$q|a_t^2 - a,$$

then from Eq. (48),

$$q|a_t^2 - a_r^2, \quad \text{or} \quad q|(a_t - a_r)(a_t + a_r).$$

Thus either $t = r$, or $a_t + a_r = mq$. In the second case, since a_t and a_r are both $<q$, $m = 1$, and therefore $a_t = q - a_r$. Thus if $\left(\frac{a}{q}\right) = +1$, there are exactly two values of a_i which satisfy the equation

$$q|x^2 - a.$$

These two values, a_r and $a_t = q - a_r$, satisfy

$$-a_r a_t = a + Kq \tag{49}$$

for some K.

The remaining $2Q - 2$ values of a_i fall into $Q - 1$ pairs (as before) and each such pair satisfies Eq. (46). The product of these $Q - 1$ equations, together with Eq. (49), gives

$$-(2Q)! = a^Q + Mq$$

for some M. Therefore

$$\left(\frac{a}{q}\right) = +1 \quad \text{implies} \quad q|(2Q)! + a^Q. \tag{50}$$

Equations (47) and (50) together read

$$q|\left(\frac{a}{q}\right)(2Q)! + a^Q. \tag{51}$$

If we let $a = 1$, by the third example of Definition 12, we have, for *every* q,

$$q|(2Q)! + 1. \tag{52}$$

Therefore $(2Q)! = -1 + Kq$ for some K, and Eq. (51) becomes

$$q|a^Q - \left(\frac{a}{q}\right). \tag{53}$$

Finally if $\left(\frac{a}{q}\right) = 0$, $q|a$, and Eq. (53) is still true. This completes the proof of Theorem 21_1.

It may be noted, that if $b^2 = a$, then by Eq. (40), and the last example of Definition 12, we again derive

$$q|b^{2Q} - 1,$$

which is Fermat's Theorem 13_1. This theorem is therefore a special case both of Euler's Theorem 14, and his Theorem 21_1.

Exercise 21. There have been many references to Fermat's Theorem in the foregoing pages. With reference to the preceding paragraph, review the proof of Theorem 21_1 to make sure that a deduction of Fermat's Theorem from Euler's Criterion is free of circular reasoning.

We have set ourselves the task of determining the odd primes $q = 2Q + 1$ which divide $a^Q - 1$. Euler's Criterion reduces that problem to the task of evaluating $\left(\frac{a}{q}\right)$. This, in turn, may be solved by *Gauss's Lemma* and the *Quadratic Reciprocity Law*. It would seem, then, that Euler's Criterion plays a key role in this difficult problem. Upon logical analysis, however, it is found to play no role whatsoever. Theorem 21 and Definition 12 will be shown to be completely unnecessary. Both are very important—for other problems. But not here. If we have nonetheless introduced Euler's Criterion at this point it is partly to show the historical development, and partly to *emphasize* its logical independence.

16. Wilson's Theorem

In the proof of Theorem 21_1 we have largely proven

Theorem 23 (Wilson's Theorem). *Let*

$$N = (q - 1)! + 1.$$

Then N is divisible by q if and only if q is a prime.

Proof (by Lagrange). The "if" follows from Eq. (52) if q is an odd prime, since $q - 1 = 2Q$. If $q = 2$, the assertion is obvious. If q is *not* a prime, let $q = rs$ with $r > 1$ and $s > 1$. Then, since $s|(q - 1)!$, $s\nmid N$. Therefore $q \nmid N$ and $q|N$ only if q is prime.

The reader will recall (page 14) that when we were still with Cataldi, we stated that a leading problem in number theory was that of finding an efficient criterion for primality. In the absence of such a criterion, we have used Fermat's Theorem 11, and Euler's Theorem 19, to alleviate the problem. Now Wilson's Theorem is a necessary and sufficient condition for primality. But the reader may easily verify that it is not a practical criterion. Thus, to prove M_{19} a prime, we would have to compute:

$$524287 | 524286! + 1. \tag{54}$$

But the arithmetic involved in Eq. (54) is much greater than even that used in Cataldi's method. We will return to this problem.

EXERCISE 22. If $q = 2Q + 1$ is prime, and Q is even,

$$q | (Q!)^2 + 1.$$

EXERCISE 23.

$$\left(\frac{-1}{q}\right) = (-1)^Q,$$

and therefore all odd divisors of $n^2 + 1$ are of the form $4m + 1$.

EXERCISE 24. For a prime $q = 4m + 1$, find all integers x which satisfy

$$q | x^2 + 1.$$

EXERCISE 25. We seek to generalize Wilson's Theorem in a manner analogous to Euler's generalization of Fermat's Theorem. Let m be an integer >1 and let a_i be the $\phi(m)$ integers $1, \cdots, m - 1$ which are prime to m. Let A be the product of these $\phi(m)$ integers a_i. Then for $m = 9$ or 10, say, we do find $m | A + 1$ analogous to $p | (p - 1)! + 1$ for p prime. But for $m = 8$ or 12 we have, instead, $m | A - 1$. Find one or more additional composites m in each of these categories. We will develop the complete theory only after a much deeper insight has been gained—see Exercise 88 on page 103.

17. Gauss's Criterion

After our digression into Euler's Criterion, we return to the problem posed on page 33. Which of the primes $q = 2Q + 1$, which do not divide a, divide $a^Q - 1$? The similarity of Theorems 18 and 20, for the cases $a = 2$ and $a = 3$, may create the impression that the problem is simpler than it really is. But consider a *larger* value of a—say $a = 17$. Then it will be found that primes of the form $34k \pm 1$, $34k \pm 9$, $34k \pm 13$, and $34k \pm 15$ divide $17^Q - 1$, while $34k \pm 3$, $34k \pm 5$, $34k \pm 7$, and $34k \pm 11$ divide $17^Q + 1$. Such complicated rules for choosing up sides seem obscure indeed. Thus the complete, and relatively simple solution for every integer a, at the hands of Euler, Legendre, and Gauss, may well be considered a Solved Problem *par excellence*.

A large step in this direction stems from the simple

Theorem 24. *Let* $a_i (i = 1, 2, \cdots, Q)$ *be the positive* odd *integers less than a prime* $q = 2Q + 1$, *and let* a *be any integer not divisible by* q. *Let*

$$aa_i = q_i q + r_i \qquad (0 < r_i < q) \tag{55}$$

as in Eq. (31) *of Theorem* 17, (*page* 23). *In addition to the result given there that all* Q *of these* r_i *are distinct, it is also true that no two of them add to* q:

$$r_i + r_j \neq q. \tag{56}$$

Proof. If $r_i + r_j = q$, then, from Eq. (55),

$$a(a_i + a_j) = Kq \tag{57}$$

for some integer K. But $a_i + a_j$ is even and $<2q$. Therefore $q \nmid a_i + a_j$ and Eq. (57) implies $q|a$. Since this cannot be, we obtain Eq. (56).

From this simple observation we obtain an important result which we will call Gauss's Criterion.

Theorem 25 (Gauss's Criterion). *Let* q, a, a_i, *and* r_i *be as in the previous theorem, and let* γ *be the number of the* r_i *which are* even—*and therefore not equal to some* a_i. *Then*

$$q|a^Q - (-1)^\gamma; \tag{58}$$

i.e., $q|a^Q - 1$ *or* $q|a^Q + 1$ *according as* γ *is even or odd.*

Proof. The set of Q remainders, r_i, given by Eq. (55), consists of γ even integers and $Q - \gamma$ of the odd integers a_i. Let each of the γ even integers, r_j, be written as $q - a_k$ for some k. But this a_k *cannot* be r_m, one of the $Q - \gamma$ odd remainders, since, if it were, we would have $r_j + r_m = q$ in violation of Theorem 24. Therefore, for each a_i, either a_i is one of the odd r_i or $q - a_i$ is one of the even r_i, but not both. In the first case we have

$$aa_j = q_j q + a_i \tag{59}$$

for some j, and in the second case we have

$$\begin{aligned} aa_k &= q_k q + (q - a_i) \\ &= (q_k + 1)q - a_i \end{aligned} \tag{60}$$

for some k.

If we now take the product of the γ equations of type Eq. (60) and the $Q - \gamma$ equations of type Eq. (59) we obtain

$$a^Q(a_1 a_2 \cdots a_Q) = Lq + (-1)^\gamma (a_1 a_2 \cdots a_Q)$$

for some integer L. Proceeding as we did in Theorem 14 (page 24) we obtain Eq. (58).

EXERCISE 26. Derive Fermat's Theorem from Gauss's Criterion, and, as in Exercise 21, check against circular reasoning.

With Gauss's Criterion we may now easily settle Theorem 18.

PROOF OF THEOREM 18. Let $a = 2$ in Theorem 25. If Q is odd, there are $(Q + 1)/2$ odd numbers, $1, 3, 5, \cdots, Q$, whose doubles

$$2\cdot 1, 2\cdot 3, \cdots, 2\cdot Q$$

are *less* than q. Therefore $q_i = 0$ in Eq. (55), and these *even* products, $2a_i$, are themselves the r_i. The remaining products

$$2(Q + 2), 2(Q + 4), \cdots, 2(2Q - 1)$$

will have $q_i = 1$ and therefore their r_i will be odd. Thus, if Q is odd, $\gamma = (Q + 1)/2$. Likewise, if Q is even, the $Q/2$ products

$$2\cdot 1, 2\cdot 3, \cdots, 2\cdot (Q - 1)$$

have even r_i, and $\gamma = Q/2$. Both cases may be combined, using Definition 6, in the formula

$$\gamma = \left[\frac{Q + 1}{2}\right] = \left[\frac{q + 1}{4}\right].$$

From Eq. (58) we therefore have

$$q|2^Q - (-1)^{[q+1/4]} \quad \text{(compare Exercise 18).} \quad (61)$$

Finally if $q = 8k \pm 1$, $\left[\frac{q + 1}{4}\right] = 2k$. And if $q = 8k \pm 3$, $\left[\frac{q + 1}{4}\right] = 2k \pm 1$. This completes the proof of Theorem 18, and therefore also of Theorem 19.

18. THE ORIGINAL LEGENDRE SYMBOL

With the proofs of Theorems 18 and 19, we might consider now whether we should pursue the general problem, $q|a^Q \pm 1$, or whether we should return quickly to the perfect numbers. But there is little occasion to do the latter. We have already remarked (page 28) that "a radically different technique is needed to go much further." Such a radically different technique is the *Lucas Criterion*. But to obtain this we need some essentially new ideas. And to prove the Lucas Criterion we will need not only Theorem 18, but also Theorem 20—the case $a = 3$. We therefore leave the perfect numbers, for now, and pursue the general problem.

Legendre's original definition of his symbol was not Definition 12, but

Definition 13 (Original Legendre Symbol). If $q = 2Q + 1$ is prime, and a is any integer, then $(a|q)$ has one of three values. If $q|a$, then $(a|q) = 0$. If not, then $(a|q) = +1$ if $q|a^Q - 1$ and $(a|q) = -1$ if $q|a^Q + 1$. In every case

$$q|a^Q - (a|q). \tag{62}$$

This *looks* very much like Euler's Criterion. But of course it isn't. It is merely a definition, not a theorem. Further, there is nothing in *this* definition about an N such that $q|N^2 - a$, etc. In view of Theorem 21_1, Eq. (40), it is clear that

$$(a|q) = \left(\frac{a}{q}\right). \tag{63}$$

We stated above, however, (page 37) that the solution of the problem $q|a^Q \pm 1$ is logically *independent* of Euler's Criterion and Definition 12. For the present then, we will ignore Eqs. (63) and (40), and confine ourselves to Definition 13 and Eq. (62).

In terms of the original Legendre symbol we may rewrite Gauss's Criterion as

Theorem 25_1. *With all symbols having their previous meaning, we have*

$$(a|q) = (-1)^\gamma \tag{64}$$

if $q \nmid a$.

The symbol $(a|q)$ has two important properties—it is *multiplicative* and *periodic*.

Theorem 26.

$$(ab|q) = (a|q)(b|q). \tag{65}$$

$$(a + kq|q) = (a|q) \qquad (\textit{for any integer } k). \tag{66}$$

Proof. Since

$$a^Q = Kq + (a|q),$$

and

$$b^Q = Lq + (b|q),$$

we have

$$(ab)^Q = Mq + (a|q)(b|q)$$

for some integers K, L, and M. But since

$$(ab)^Q = Nq + (ab|q)$$

we have

$$q|(a|q)(b|q) - (ab|q). \tag{67}$$

Again, by Eq. (13),

$$q|(a + kq)^n - a^n$$

for every n. Therefore $q|(a + kq)^Q - a^Q$ or

$$q|(a + kq|q) - (a|q). \tag{68}$$

Since the right sides of Eqs. (67) and (68) are less than q in magnitude, they must both vanish, and therefore Eqs. (65) and (66) are true.

To solve the problem $q|a^Q \pm 1$, we must evaluate $(a|q)$. If $q|a$, there is no problem. Let $q\nmid a$ and let a be a positive or negative integer written in a standard form

$$a = \pm p_1^{a_1} p_2^{a_2} \cdots p_n^{a_n}. \tag{69}$$

By factoring out $p_i^{a_i}$ for every *even* a_i, and $p_j^{a_j-1}$ for every odd $a_j > 1$, we are left with

$$a = \pm p_j p_k \cdots p_m N^2, \tag{70}$$

a product of primes times a perfect square N^2. Now, from Eq. (65),

$$(N^2|q) = (N|q)(N|q) = +1 \tag{71}$$

since $q\nmid N$, so that, from Eq. (65), we have

$$(a|q) = (\pm 1|q)(p_j|q) \cdots (p_m|q). \tag{72}$$

If the first factor is $(-1|q)$, from Eq. (62) we have

$$(-1|q) = (-1)^Q. \tag{73}$$

Otherwise

$$(1|q) = 1. \tag{74}$$

If $p_j = 2$, from Eq. (61) we have

$$(2|q) = (-1)^{[(q+1)/4]}. \tag{75}$$

Therefore to evaluate any $(a|q)$ there remains the problem of evaluating $(p|q)$ for any two odd primes p and q.

19. The Reciprocity Law

By examining many empirical results (such as those of Exercise 19), Euler, Legendre and Gauss independently discovered a most important theorem. But only Gauss proved it—and it took him a year. In terms of the original Legendre Symbol we write

Theorem 27 (The Reciprocity Law). *If* $p = 2P + 1$ *and* $q = 2Q + 1$ *are unequal primes, then*

$$(p|q)(q|p) = (-1)^{PQ}. \tag{76}$$

The theorem may also be stated as follows: $(p|q) = (q|p)$ unless p and q are both of the form $4m + 3$. In that case, PQ is odd, and $(p|q) = -(q|p)$.

Before we prove Theorem 27, let us state right off that it completely solves our problem, $q|a^Q \pm 1$. We stated above that what remained was to evaluate all $(p|q)$. But if $p > q$ we may write $p = sq + r$ and therefore, by Eq. (66), $(p|q) = (r|q)$. Without loss of generality we may therefore assume $p < q$. But in that case we may use Eq. (76) and obtain

$$(p|q) = (-1)^{PQ}(q|p)$$

so that if $q = sp + r$,

$$(p|q) = (-1)^{PQ}(r|p).$$

Thus we reduce a symbol whose right argument is a prime q to one whose right argument is a *smaller* prime p. By continuing this reduction we must eventually get down to a symbol

$$(-1|q), \quad (1|q), \quad \text{or} \quad (2|q)$$

which we can evaluate by Eqs. (73), (74) or (75).

To illustrate these reductions we will evaluate several $(a|q)$ and prove one theorem. In carrying out any step of a reduction it will be convenient to write

$$\begin{aligned}
(1|q) &= 1_U \\
(-1|q) &= (-1)_N^Q \\
(2|q) &= (-1)_D^{[(q+1)/4]} \\
(aN^2|q) &= (a|q)_S \ (\text{if } q \nmid N) \\
(ab|q) &= (a|q)(b|q)_M \\
(a + kq|q) &= (a|q)_P \\
(p|q) &= (-1)^{PQ}(q|p)_R
\end{aligned} \tag{77}$$

depending on whether that step uses the "unit," "negative," "double," "square," "multiplicative," "periodic," or "reciprocity" rule. There is no unique method of reduction. Thus

$$(8|17) = (2|17)_S = +1_D,$$

or

$$(8|17) = (-9|17)_P = (-1|17)_S = +1_N,$$

or

$$(8|17) = (25|17)_P = (1|17)_S = +1_U,$$

or

$$(8|17) = (-26|17)_P = (26|17)_{MN} = (13|17)_{MD}$$
$$= (17|13)_R = (4|13)_P = +1_{SU}.$$

Any path leads to the same answer, $(8|17) = 1$, and this implies $17|8^8 - 1$.

Again,

$$(17|47) = (47|17)_R = (13|17)_P = +1 \qquad \text{(as above)}.$$
$$(17|47) = (64|17)_P = +1_{SU}.$$

Therefore $47|17^{23} - 1$.

A third example raises a new point. We have

$$(15|47) = (3|47)(5|47)_M = -1(47|3)(47|5)_R$$
$$= -1(-1|3)(2|5)_P = (2|5)_N = -1_D.$$

$$\text{But } (15|47) \overset{?}{=} -1(47|15)_R \overset{?}{=} -1(2|15)_P \overset{?}{=} -1_D.$$

In the second "reduction" we did not factor 15 and we applied the rules (77) to $(47|15)$ and $(2|15)$. But 15 is not a prime! *Nonetheless* we obtained the correct answer. We will return to this pleasant possibility in Volume II when we study the *Jacobi Symbol*. Now let us prove Theorem 20.

PROOF OF THEOREM 20.

$$(3|12k + 1) = (12k + 1|3)_R = (1|3)_P = 1_U.$$
$$(3|12k - 1) = -1\cdot(12k - 1|3)_R = -1\cdot(-1|3)_P = 1_N.$$
$$(3|12k + 5) = (12k + 5|3)_R = (-1|3)_P = -1_N.$$
$$(3|12k - 5) = -1(12k - 5|3)_R = -1\cdot(1|3)_P = -1_U.$$

Therefore $\qquad q = 12k \pm 1|3^Q - 1 \quad$ and

$$q = 12k \pm 5|3^Q + 1.$$

We note, in passing, that Theorem 20 makes an assertion concerning $(3|q)$ for infinitely many q, while in the proof we need evaluate only finitely many Legendre Symbols. It is, of course, the Reciprocity Law, together with Eq. (66), that brings about this economy.

EXERCISE 27. Verify the statements on page 38 concerning $17^Q \pm 1$.

EXERCISE 28. Investigate the possibility of always avoiding the "double" rule, inasmuch as

$$(2|q) = (-1|q)(q - 2|q)$$

If so, it means that our original motivation, $q|2^Q \pm 1$, is the *one* thing we do *not* need in determining $q|a^Q \pm 1$.

The simplest and most direct proof of the Reciprocity Law is perhaps the following modification of a proof by Frobenius. It is based on Gauss's Criterion.

PROOF OF THEOREM 27. Let $q = 2Q + 1$ and $p = 2P + 1$ be distinct primes. Let a be an *odd* integer satisfying $0 < a < p$ such that

$$qa = pa' + r \tag{78}$$

with r an *even* integer satisfying $0 < r < p$. If γ is the number of such a, by Eq. (64) we have

$$(q|p) = (-1)^{\gamma}.$$

It follows from Eq. (78) that for each such a, the corresponding a' is also *odd*, is unique, and satisfies $0 < a' < q$.

By symmetry

$$(p|q) = (-1)^{\gamma'}$$

where γ' is the number of odd a' such that

$$pa' = qa + r \tag{79}$$

with $0 < a' < q$, $0 < r < q$, $0 < a < p$, and with a odd. Again, for each such a', a is unique.

If we now consider the function

$$R(a, a') = qa - pa'$$

where $a = 1, 3, 5, \cdots, p - 2$, and $a' = 1, 3, 5, \cdots, q - 2$, we see that there are γ of these R which satisfy $0 < R < p$, and γ' of these R which satisfy $-q < R < 0$. Since there is no $R = 0$ (because $a < p$ and p is a prime), we see that there are $\gamma + \gamma'$ values of R such that

$$-q < R < p. \tag{80}$$

But if $R_1 = qa_1 - pa_1'$ is one of these, then so is $R_2 = qa_2 - pa_2'$ where

$$\begin{aligned} a_1 + a_2 &= p - 1, \\ a_1' + a_2' &= q - 1, \end{aligned} \tag{81}$$

and therefore

$$R_1 + R_2 = p - q.$$

For, the mean value of R_1 and R_2 equals the mean value of the limits of Eq. (80), $-q$ and p. Therefore if R_1 is even, and between these limits, so will R_2 be even and between the limits. And likewise if a_1 is odd and between 0 and $p - 1$, so is a_2. And similarly with a_1' and a_2'.

Therefore each R in Eq. (80) has a companion R in Eq. (80), given by Eq. (81), *unless*

$$a_1 = a_2 = (p - 1)/2 = P, \quad \text{and} \quad a_1' = a_2' = Q. \tag{82}$$

But, since every a and a' is odd, Eq. (82) cannot occur unless P and Q are both odd. Conversely, if P and Q are both odd, there is a self-companioned

$$R = qP - pQ = P - Q$$

given by Eq. (82), which does satisfy Eq. (80).

Thus $\gamma + \gamma'$ is *even* unless P and Q are both *odd*. But so is PQ even, unless P and Q are both odd. Therefore

$$(q|p)(p|q) = (-1)^{\gamma+\gamma'} = (-1)^{PQ}$$

and Theorem 27 is proven.

Gauss gave seven or eight different proofs of the Reciprocity Law. All of them were substantially more complicated than the one we have given—and the first proof, as we have said above, took him a year to obtain. Yet the given proof, based on Gauss's Criterion, seems quite straightforward and simple. We will return later to this question—since we are interested, among other things, in the reasons why some proofs are complicated, and in the feasibility of simplifying them.

We may note that the proofs of Theorem 25 (Gauss's Criterion) and of Theorem 27 just given, are similar in *strategy* to parts of Dirichlet's proof of Euler's Criterion (page 35). In both cases we multiply Q equations together, and in both cases we set up "companions"—except that in Euler's Criterion the companions are multiplicative, as in Eq. (46), while in Theorem 27 they are additive, as in Eq. (81). Again, in both cases, the self-companioned singularity (which may or may not occur) is the critical point of the proof.

Exercise 29. Show that if the Q numbers a_i in Theorem 24 are the numbers $1, 2, \cdots, Q$ instead of the odd numbers, the theorem is still true.

Exercise 30. Modify Theorem 25 in accordance with the different set of a_i in the previous exercise. (For this different set and with the use of $\left(\frac{a}{q}\right)$ instead of $(a|q)$, this result is called Gauss's Lemma.) Carry out the details of the new proof.

Exercise 31. With the variation on Theorem 25 of the previous example, carry out another proof of Theorem 27—with "companions," etc.

Exercise 32. Consider Eq. (80) and show that each R such that $p < R$ can be put into one-to-one correspondence with an R such that $R < -q$.

If the number in each set is A, then $PQ = \gamma + \gamma' + 2A$. Therefore we have another variation on the proof of Theorem 27.

EXERCISE 33. Examine the "companions," Eq. (81), in several numerical cases and verify that sometimes the γ solutions of Eq. (78) choose their companions solely from the γ' solutions of Eq. (79), while sometimes some of the γ companions of Eq. (78) are themselves from the set Eq. (78).

20. THE PRIME DIVISORS OF $n^2 + a$

Now that we have completed the solution of the problem $q|a^Q \pm 1$, we will lift our ban against Euler's Criterion and Definition 12. Henceforth, $(a|q)$ and $\left(\frac{a}{q}\right)$ are identical, will be designated *the* Legendre Symbol, and may be written in either notation.

If $q = 2Q + 1$ is a prime which does not divide a, we now have at once that

$$q|n^2 + a$$

for *some* n, if $\left(\frac{-a}{q}\right) = +1$, and

$$q \nmid n^2 + a$$

for *any* n, if $\left(\frac{-a}{q}\right) = -1$. The symbol $\left(\frac{-a}{q}\right)$ we may evaluate by the rules of Eq. (77). When we are concerned, as we are here, with $q|n^2 + a$, Theorem 27 is called the *Quadratic Reciprocity Law*.

EXERCISE 34. Show that for any odd prime $q = 24k + b$, which is greater than 3, there is an n such that $q|n^2 + a$ if b is one of those listed in the following table. If b is not so listed, in the row corresponding to a, there is no such n.

a	b
$+1$	1, 5, 13, 17
$+2$	1, 11, 17, 19
-2	1, 7, 17, 23
$+3$	1, 7, 13, 19
-3	1, 11, 13, 23
$+6$	1, 5, 7, 11
-6	1, 5, 19, 23

COMMENT: In each of these seven cases "one-half" of the primes divide the numbers of the form $n^2 + a$, since $\phi(24) = 8$. (When we get to *modulo multiplication groups*, these seven sets of b will constitute the seven *subgroups* of order four in the group modulo 24. Why the special role of $b = 1$? Because 1 is the identity element of the group.)

EXERCISE 35. Prove the conjecture you made concerning the prime divisors of $n^2 - 5$ in Exercise 19. Or, if your conjecture was erroneous, disprove it. But if you haven't done Exercise 19, don't do it now. You already know too much.

The reader no doubt asked himself, while reading Conjectures 11 and 12, why there should be more primes of the form $n^2 - 2$ than of the form $n^2 + 1$, and what the general situation would be for any form $n^2 + a$. With what he knows now the reader may begin, if he wishes, to partially formulate his own answer. In particular, from the table in Exercise 34, should there be (relatively) few primes of the form $n^2 + 6$, or (relatively) many?

Definition 14. By $P_a(N)$ is meant the number of primes of the form $n^2 + a$ for $1 \leqq n \leqq N$. If a is negative, and if for some n, $n^2 + a$ is the negative of a prime, we will, nonetheless, count it as a prime.

Now that we have the Legendre symbol we can define the constants in Conjectures 11 and 12, and state a general conjecture of which these two are special cases.

Conjecture 12_1 (Hardy-Littlewood).
If $a \neq -m^2$,

$$P_a(N) \sim \frac{1}{2} h_a \int_2^N \frac{dn}{\log n}$$

where the constant h_a is given by the infinite product

$$h_a = \prod_w^{\infty} \left\{ 1 - (-a|w) \frac{1}{w - 1} \right\}$$

taken over all odd primes w. Here $(-a|w)$ is the Legendre symbol.

EXAMPLE:
From $(-1|w) = (-1)^{(w-1)/2}$ we have

$$h_1 = (1 + \tfrac{1}{2})(1 - \tfrac{1}{4})(1 + \tfrac{1}{6})(1 + \tfrac{1}{10})(1 - \tfrac{1}{12})(1 - \tfrac{1}{16}) \cdots$$
$$= 1.37281346 \cdots,$$

and we thus obtain Eq. (37) for primes of the form $n^2 + 1$.

But to evaluate such slowly convergent infinite products we will need many things which we have not yet developed—Möbius Inversion Formula,

Gauss Sums, and Dirichlet Series. We therefore postpone further consideration of this conjecture until Volume II.

We offer, however, without further comment, a little table for the reader's consideration.

a	$P_a(1000)$	$P_a(10{,}000)$	$P_a(100{,}000)$
+7	167	1238	9521
−2	157	1153	8888
−5	148	1088	8579
−3	120	850	6664
+1	112	841	6656
+4	125	870	6517
+3	109	711	5426
−6	91	643	5010
−7	68	440	3627
+2	68	446	3422
+6	53	444	3420
+5	48	339	2567
−4	2	2	2
−1	1	1	1
0	0	0	0

tion, with an unsubstantial cover, and with pages so well oxidized that it may well attain this "Dirichlet Condition" even if it encounters a more casual reader. There also exists a German translation((1889), but, at this writing, the book is still not available in English.

We ask now, what was in it; and why did it make such a splash? Well, many *new* things were in it—Gauss's proof of the Reciprocity Law, his extensive theory of binary quadratic forms, a complete treatment of primitive roots, indices, etc. Finally it included his most astonishing discovery, that a regular polygon of $F_m = 2^{2^m} + 1$ sides can be inscribed in a circle with a ruler and compass—provided F_m is a prime.

But the most immediate thing found in Gauss's book was not one of these new things; it was a new way of looking at the old things. By this new way we mean the *residue classes*. Gauss begins on page 1 as follows:

"If a number A divides the difference of two numbers B and C, B and C are called *congruent* with respect to A, and if not, *incongruent.* A is called the *modulus*; each of the numbers B and C are *residues* of each other in the first case, and *non-residues* in the second."

Does it seem strange that Gauss should write a whole book about the implications of

$$A|B - C \text{ ?} \tag{83}$$

It surely is not clear *a priori* why Eq. (83) should be worthy of such protracted attention. In fact, these opening sentences are completely unmotivated, and hardly understandable, except in the historical light of the previous chapter. But in that light, the time was ripe—and even overripe—for such an investigation. We will review four aspects of the situation then existing.

(a) First, it will not have escaped the reader that we were practically surrounded by special instances of (83) in the previous chapter. Thus Fermat's Theorem 13_1 reads:

$$p|a^{p-1} - 1,$$

and his Theorem 11:

$$q|2^p - 1 \rightarrow 2p|q - 1,$$

can go it one better by having both hypothesis and conclusion in that form. So likewise Euler's Criterion:

$$q|N^2 - a \leftrightarrow q|a^Q - 1,$$

and his Theorem 19:

$$q|2^p - 1 \rightarrow 8|q - 1 \quad \text{or} \quad 8|q - (-1).$$

Could so much formal similarity be fortuitous? And if not, what could be its significance?

Where we *first* came upon such expressions we know well enough—if $N = 2^{n-1}F$ is to be perfect, the sum of divisors $1 + 2 + \cdots + 2^{n-1} = 2^n - 1$ must be a divisor of N, and must also be a prime. But $23|2^{11} - 1$, and therefore M_{11} was not a prime, etc. It is another question, however, if we ask why the expressions $A|B - C$ should be so *persistent.*

We should make it clear, at this point, that though we have followed one path in the previous chapter, that starting from the perfect numbers, much other ground had been gone over by this time. In particular, consider Gauss. Gauss could compute as soon as he could talk—in fact, he jokingly claimed he could compute even earlier. He rediscovered many of the theorems given in the previous chapter before he had even heard of Fermat, Euler or Lagrange. It is clear that no computing child could *reinvent* anything as esoteric as the perfect numbers, and therefore Gauss could not have followed the path which we have sketched. To the Greeks a divisor of a number, other than itself, was a "part" of the number; and for a perfect number, the whole was equal to the sum of its parts. Such a Greek near-pun could well engage the classicists of the Renaissance, but would not be likely to occur to a self-taught Wunderkind.

What was available to Gauss was such material as

$$\tfrac{1}{7} = .142857142857142857 \cdots$$

and

$$\tfrac{1}{13} = .076923076923076923 \cdots .$$

Now if $\frac{1}{7}$ is a periodic decimal with a period 6, then since $1 = 0.999999 \cdots$, it means that $7|999999$, or

$$7|10^6 - 1.$$

Likewise for any prime p, not equal to 2 or 5, we find

$$p|10^{p-1} - 1.$$

Therefore, say,

$$13|10^{12} - 1.$$

But we have just seen that $\frac{1}{13}$ also has a period of 6, so that

$$13|10^6 - 1.$$

From the foregoing theory we know that

$$p|10^{(p-1)/2} - 1 \quad \text{means} \quad (10|p) = +1$$

and that

$$(10|13) = (-3|13)_P = (3|13)_{MN} = (13|3)_R = +1_{PU}.$$

It is clear, however, that whether Fermat and Euler were interested in perfect numbers—and $23|2^{11} - 1$; or Gauss was interested in periodic decimals—and $13|10^6 - 1$, the basic underlying theorems are identical, and $A|B - C$ arises in either case.

(b) There is another case of persistence in the previous chapter. On pages 24, 27, 35, etc., we are saying, repeatedly, "for *some* integer, Q, L, K, K_2" etc., and that seems almost paradoxical at first. Isn't number theory an exact science—don't we *care* what Q, L, etc., are equal to? The answer is, generally,* no. If we are interested in $A|B$ this implies *some* integer X such that $B = AX$, but *which* integer is quite irrelevant.

It is instructive to examine the *additive* analogue of divisibility, $A < B$. This implies a positive X such that $B = A + X$, but which X is again irrelevant. If this were not the case, Analysis would be quite impossible. It is difficult enough to show that a certain quantity is *less* than epsilon—it would be totally unfeasible if we always had to tell *how much* less. The analyst embodies this *ambiguity* in X by working with classes of numbers, $-\epsilon < X < \epsilon$, and *any* X in the class will do. Likewise in divisibility theory we should consider the advantages of working with classes of numbers, which would embody the ambiguity presently in question.

A variation on this theme concerns the *algebra* of such ambiguity. On page 27 we square one ambiguous equation, $2 = N^2 - Kq$, to obtain a second, $2^2 = N^4 - K_2q$. On page 36 we substitute the ambiguous $N = sq + a_r$ into $q|N^2 - a$ to obtain $q|a_r^2 - a$. Such persistent, redundant, and rather clumsy algebra virtually demands a new notation and a new algebra.

(c) Again, consider the *arithmetic* of page 26:

$$167|2^{83} - 1,$$

or the seemingly impossible operation,

$$32070004559|2^{16035002279} - 1,$$

of Exercise 7. The first seems a little long and the second virtually impossible—but only because the dividend, and therefore the *quotient* is so large. But we said that in questions of divisibility the quotient is *irrelevant*, that only the remainder is of importance. Thus, if

$$b = qa + r,$$

divisibility depends only on r. And r is less than a. And a, even in the second case, is not too large to handle. What we want, then, is an arithmetic of remainders.

---

* An important exception will be discussed in Sect. 25.

(d) A final, and most important point. Fermat's Theorem quickly let its power be seen. Thus

$$f_{19} = 6 \ll c_{19} = 128$$

was most impressive. Similarly Euclid's Theorem 5 and its immediate consequence Theorem 6 have, by their constant use, become quite indispensable. Yet can we say, at this point, that we can see clearly the source of this power and this indispensability? There is suggested here the existence of a deeper, underlying structure, the investigation of which deserves our attention.

We want then, in (b), an algebra of ambiguity; in (c), an arithmetic of remainders; and in (d), an interpretation in terms of an underlying structure. It is the merit of the *residue classes* that they answer all three of these demands.

We could, it is true, have introduced them earlier—and saved a line here and there in the proofs. But History did *not* introduce them earlier. Nor would it be in keeping with our title, "Solved and Unsolved Problems," for us to do so. To have a solved problem, there must first be a *problem*, and *then* a solution. We could not expect the reader to appreciate the solution if he did not already appreciate the problem. Moreover, if we have gone on at some length before raising the curtain (and perhaps given undue attention to lighting and orchestration) it is because we thought it a matter of some importance to analyze those considerations which may have led Gauss to invent the residue classes. Knowing what we do of Gauss's great skill with numbers, and while we can not say for certain, the consideration most likely to have been the immediate cause of the invention would seem to be item (c) above.

Exercise 36. Using the results of Exercise 35 and of Exercise 18, determine the odd primes $p = 2P + 1 \neq 5$ such that $1/p$ has a decimal expansion which repeats every P digits. The *period* of some of these primes may be less. Thus $\frac{1}{37} = .027027 \cdots$ does repeat every 18 digits, but its period is 3.

22. The Residue Classes as a Tool

Definition 15. If a, b, and c are integers, with $a > 0$, and such that

$$a|b - c, \tag{84}$$

we may write, equivalently,

$$b \equiv c \pmod{a}, \tag{85}$$

and the latter is read "b is *congruent to* c *modulo* a." We may also say "b is a *residue of* c modulo a." Conversely, given Eq. (85), we may write Eq.

(84). If b is not congruent to c modulo a, we write

$$b \not\equiv c \pmod{a}. \tag{86}$$

If

$$b = qa + r, \tag{87}$$

then

$$b \equiv r \pmod{a}$$

independently of the value of q. As q takes on all integral values, $\cdots, -2, -1, 0, 1, 2, \cdots$, each such b is congruent to r, and *all* such b form a class of numbers which we call a *residue class*. a is called the *modulus*.

EXAMPLES:

$$2^{11} \equiv 1 \pmod{23}.$$
$$2^{29} \equiv -1 \pmod{59}.$$

(Fermat's Theorem)

$$a \not\equiv 0 \pmod{p} \to a^{p-1} \equiv 1 \pmod{p}.$$

(Euler's Criterion)

$$N^2 \equiv a \pmod{p} \leftrightarrow a^{(p-1)/2} \equiv 1 \pmod{p}.$$

For any $a > 0$, and any b we can always write Eq. (87) with $0 \leqq r < a$. Corresponding to a modulus a, there are therefore a distinct residue classes, and the integers $0, 1, 2, \cdots, a - 1$ belong to these distinct classes, and may be used as names for these classes. Thus we may say 35 belongs to residue class 3 modulo 16.

"Congruent to" is an *equivalence relation*, in that all three characteristics of such a relation are satisfied. Specifically:

Reflexive. For all b, (88*a*)

$$b \equiv b \pmod{a}.$$

Symmetric. (88*b*)

$$b \equiv c \pmod{a} \text{ implies } c \equiv b \pmod{a}.$$

Transitive. (88*c*)

$$b \equiv c \pmod{a} \text{ and } c \equiv d \pmod{a}$$
$$\text{implies } b \equiv d \pmod{a}.$$

All the numbers in a residue class are therefore congruent to each other (mod a).

The utility of residue classes comes from the fact that this equivalence is preserved under addition, subtraction and multiplication. Thus we have

Theorem 28. *Let* $f(a, b, c, \cdots)$ *be a* polynomial *in* r *variables with integer*

coefficients. That is, f is a sum of a finite number of terms, $na^{\alpha}b^{\beta}\cdots$, *each being a multiple of a product of powers of the variables. Here n is an integer and* $\alpha, \beta, \cdots$ *are nonnegative integers. If* $a_1, b_1, c_1, \cdots$ *are integers, and if*

$$N_1 = f(a_1, b_1, c_1, \cdots), \tag{89}$$

and if

$$a_1 \equiv a_2, \quad b_1 \equiv b_2, \quad c_1 \equiv c_2, \cdots \quad (\text{mod } M) \tag{90}$$

then

$$N_2 = f(a_2, b_2, c_2, \cdots) \equiv N_1 \quad (\text{mod } M). \tag{91}$$

Proof. The reader may easily verify that if Eq. (90) is true, then so are

$$\begin{aligned} a_1 + b_1 &\equiv a_2 + b_2 \quad (\text{mod } M), \\ a_1 - b_1 &\equiv a_2 - b_2 \quad (\text{mod } M), \\ a_1 b_1 &\equiv a_2 b_2 \quad (\text{mod } M). \end{aligned} \tag{92}$$

By induction, it is clear that any finite number of these three operations may be compounded without changing the residue class, and since any polynomial, Eq. (89), may be thus constructed, the theorem is true.

Corollary. *If* $f(a)$ *is a polynomial in one variable, then*

$$a \equiv a' \quad (\text{mod } M) \quad \textit{implies} \quad f(a) \equiv f(a') \quad (\text{mod } M).$$

This simple theorem allows us to use the residue classes as a tool for those arithmetic and algebraic problems which we discussed on page 54.

Consider some simple examples.

(a) To verify that $7 | 10^6 - 1$, we may write

$$10^6 \equiv 3^6 \quad (\text{mod } 7)$$

since $10 \equiv 3$. But

$$3^2 \equiv 2 \quad \text{and thus} \quad 10^6 \equiv 3^6 \equiv 2^3 \equiv 1 \quad (\text{mod } 7).$$

Therefore

$$7 | 10^6 - 1.$$

(b) To determine if 167 divides M_{83}, we may proceed as follows:

$$2^8 = 256 \equiv 89 \quad (\text{mod } 167)$$

$$\therefore 2^{16} \equiv 89^2 = 7921 \equiv 72 \quad (\text{mod } 167)$$

$$\therefore 2^{32} \equiv 72^2 = 5184 \equiv 7 \quad (\text{mod } 167)$$

$$\therefore 2^{64} \equiv 49, \quad \text{and} \quad 2^{67} \equiv 49 \cdot 8 \equiv 58 \quad (\text{mod } 167)$$

$$\therefore 2^{83} = 2^{67} \cdot 2^{16} \equiv 58 \cdot 72 \equiv 4176 \equiv 1 \quad (\text{mod } 167)$$

$$\therefore 167 | 2^{83} - 1.$$

The advantage of the congruence notation is clear. What we really want to know here is whether 2^{83} and 1 are in the *same* residue class, and in our computation of 2^{83} we continually reduce the partial results to *smaller* members of the residue class, thus keeping the numbers from becoming unduly large.

(c) Aside from advantages in the *computation* of results, there is also an advantage in their *presentation*. Thus to show that $641 | 2^{32} + 1$, the presentation

$$641 \overline{)4294967297}^{\;6700417}$$

lacks the property of being easily checked mentally. But consider

$$640 = 5 \cdot 128 = 5 \cdot 2^7 \equiv -1 \pmod{641}.$$

$$\therefore 5^4 \cdot 2^{28} \equiv 1 \pmod{641}.$$

But $$5^4 = 625 \equiv -16 = -2^4 \pmod{641}.$$

$$\therefore -2^{32} \equiv 1 \pmod{641}$$

or $$2^{32} + 1 \equiv 0 \pmod{641}.$$

Here the arithmetic is easily verified mentally.

(d) The proofs of some of the theorems in the previous chapter could have been written more compactly in the new notation. For example, on page 27, if $q | N^2 - 2$, then

$$N^2 \equiv 2 \pmod{q}$$

and *directly* we may write

$$2^Q \equiv N^{2Q} \equiv 1 \pmod{q}.$$

Thus by setting up an algebra of ambiguity (page 55) we have simultaneously rid ourselves of the "some integer K" (page 27) which is clearly redundant and merely extends the computation.

But to complete our algebraic tools we need division also, and for this we have

Theorem 29 (Cancellation Law). *If* $bc \equiv bd \pmod{a}$ *and* $(b, a) = 1$ *then* $c \equiv d \pmod{a}$.

This is only a restatement of Theorem 6 in the new notation. We will reprove it using this notation.

PROOF. If $(b, a) = 1$, from Eq. (7), page 9, we have

$$nb \equiv 1 \pmod{a}. \tag{93}$$

Therefore if

$$bc \equiv bd, \qquad nbc \equiv nbd, \qquad \text{or} \quad c \equiv d \qquad (\text{mod } a).$$

Equation (93) is the key to our next topic, the Residue Classes as a *Group*.

EXERCISE 37. Prove Theorem 22, page 35, and Theorem 21_1, page 35, in the congruence notation.

EXERCISE 38. Verify that

$$1823 | M_{911} .$$

23. THE RESIDUE CLASSES AS A GROUP

In the previous sections the integers were the sole objects of our attention, and, as long as we considered the residue classes merely as a tool, this remained the case. We now consider a system of residue classes as a mathematical object in its own right, and, in particular, we study the multiplicative relationships among these classes.

For a modulus m there are m residue classes, which we designate $0, 1, \cdots, m - 1$, the ath class being that which contains the integer a. The system of these m classes is therefore not infinite, like the integers, but is a finite system with m elements. By the *product* of two classes a and b we mean the class of all products $a_1 b_1$ where

$$a_1 \equiv a \quad \text{and} \quad b_1 \equiv b \qquad (\text{mod } m).$$

By Eq. (92) all these products lie in a residue class, say c, and we write

$$ab \equiv c \qquad (\text{mod } m).$$

For example, for $m = 7$, we have the following multiplication table:

$a \backslash b$	0	1	2	3	4	5	6
0	0	0	0	0	0	0	0
1	0	1	2	3	4	5	6
2	0	2	4	6	1	3	5
3	0	3	6	2	5	1	4
4	0	4	1	5	2	6	3
5	0	5	3	1	6	4	2
6	0	6	5	4	3	2	1

$$ab \equiv c \qquad (\text{mod } 7).$$

If $(a, m) = 1$ and $a \equiv a_1 \pmod{m}$, we have $(a_1, m) = 1$. Thus we may say that the *residue class* a is *prime to* m. Now if $(a, m) = 1$ we have an a' and m' such that

$$a'a + m'm = 1 \tag{94}$$

and conversely. Therefore

$$a'a \equiv 1 \pmod{m}. \tag{95}$$

Definition 16. We may call the a' and a in Eq. (95) the *reciprocals* of each other modulo m, and write

$$a^{-1} \equiv a' \pmod{m}. \tag{96}$$

We may therefore characterize the $\phi(m)$ residue classes prime to m as those which possess reciprocals. If $(a, m) = (b, m) = 1$, then so is $(ab, m) = 1$, by Theorem 5, Corollary. In fact, since

$$a^{-1}ab^{-1}b \equiv 1 \pmod{m},$$

we have explicitly

$$(ab)^{-1} \equiv a^{-1}b^{-1} \pmod{m}. \tag{97}$$

We will have occasion, say in Eqs. (103a) and (104a) on page 66, and in Eq. (136) on page 100, to calculate the reciprocal of a modulo m. This we do by obtaining Eq. (94) from Euclid's Algorithm as on page 9. Equivalently, one may utilize the continued fraction (12) on page 12 with the term $1/q_n$ omitted. This fraction we evaluate by the method on page 183 below. The denominator so obtained, or its negative, is the reciprocal of a modulo b. This follows from the analogue of Eq. (271).

Definition 17. A *group* is a set of elements upon which there is defined a binary operation called *multiplication* which

(A) is *closed*, that is, if

$$c = ab,$$

then c is in the group if a and b are; and

(B) is *associative*, that is,

$$(ab)c = a(bc)$$

for every a, b and c.

Further,

(C) the group possesses an *identity* element (write it 1) such that

$$1a = a$$

for every a; and also

(D) it possesses *inverse* elements (write these a^{-1}) such that

$$a^{-1}a = 1$$

for every a.

Thus the $\phi(m)$ residue classes prime to m form a group under the binary operation multiplication modulo m. The postulates (B) and (C) are trivially true, while closure (A), from Eq. (97), and inverses (D), from Eq. (96), both stem from Eq. (94), that is, from Euclid's Theorem 5.

Definition 18. If the operation in a group is commutative, that is, if

$$ab = ba$$

for each a and b, the group is called *Abelian*. If the number of elements in a group is finite, the group is *finite*, and the *order* of the group is the number of elements.

Definition 19. The group of $\phi(m)$ residue classes prime to m, under multiplication modulo m, we call a *modulo multiplication group*, and we write it $\mathfrak{M}_m$. It is a finite, Abelian group of order $\phi(m)$.

The theory of finite groups is a large subject, into which we shall scarcely enter. We shall confine ourselves primarily to $\mathfrak{M}_m$. Nonetheless, there is a value here in introducing the more abstract Definition 17, and that lies in the *economy* of this definition. In any theorem, say for $\mathfrak{M}_m$, which we deduce from these four postulates, we have a certain assurance that redundancies and irrelevancies have not entered into the proof. Pontrjagin puts it this way:

"The theory of abstract groups investigates an algebraic operation in its purest aspect."

Several of our foregoing theorems have a simple group-theoretic interpretation. We will illustrate them using the multiplication table for $\mathfrak{M}_7$.

1	2	3	4	5	6
2	4	6	1	3	5
3	6	2	5	1	4
4	1	5	2	6	3
5	3	1	6	4	2
6	5	4	3	2	1

(Note that the row and column headings are omitted, since the first row and column also serve this purpose.)

Theorem 17 says that if

$$aa_i \equiv r_i \pmod 7$$

the r_i are a permutation of the a_i—that is, each *row* in the table contains every element. But this is true for *every* finite group.

Again, Theorem 22 says that

$$xa_i \equiv a \pmod 7$$

has a unique solution—that is, each *column* in the table contains every element. Again, this is true for every finite group.

Since in an Abelian group the rows and columns are identical, we now realize that Theorem 22 is essentially a restatement of Theorem 17. We have seen previously that Fermat's Theorem 13 may be deduced either from Euler's Theorem 14, or from Euler's Theorem 21_1, and we now note that the corresponding underlying Theorems 17 and 22 are also equivalent.

Euler's Theorem 14 says that $(a, 7) = 1$ implies

$$a^6 \equiv 1 \pmod 7.$$

Again, for every group of order n, $a^n = 1$ is valid for every element a. In fact, the whole subject of finite group theory may be thought of as a generalization of the theory of the roots of unity. It is not surprising, then, that Fermat's Theorem plays such a leading role, seeing, as we now do, that it merely expresses the basic nature of any finite group.

The three theorems just discussed hold for $\mathfrak{M}_m$ whether m is a prime or not. But Euler's Criterion does not generalize so simply. This criterion states that

$$a^{\phi(p)/2} \equiv 1 \pmod p \leftrightarrow n^2 \equiv a \pmod p. \tag{98}$$

But consider $m = 8$ and $m = 10$. In both cases $\frac{1}{2}\phi(m) = 2$. Now for the modulus $m = 10$, the implication (98) still holds. But for $m = 8$, we have

$$3^2 \equiv 1 \pmod 8$$

while

$$n^2 \equiv 3 \pmod 8$$

has no solution. This is a difference which we shall investigate. It is associated with a particular characterization of the $\mathfrak{M}_m$ groups for every m which is prime, and for *some* m which are composite; namely, that these groups have a property which we shall call *cyclic*.

Exercise 39. Write out the multiplication tables for $\mathfrak{M}_9$ and $\mathfrak{M}_{14}$. (If you use the commutative law, and the generalized Theorems 17 and 22 mentioned above, you will save some arithmetic.)

EXERCISE 40. If $(a, m) = 1$, show that

$$a^{-1} \equiv a^{\phi(m)-1} \pmod{m}. \tag{99}$$

Further, if $(a, m) = g$, $a = \alpha g$, and $m = \mu g$,

then $$a' = \alpha^{\phi(\mu)-1}$$

and $$m' = (1 - \alpha^{\phi(\mu)})/\mu$$

are integers that satisfy

$$a'a + m'm = g.$$

24. QUADRATIC RESIDUES

Definition 20. Any residue class lying on the principal diagonal of the $\mathfrak{M}_m$ multiplication table is called a *quadratic residue* of m. That is, a is a quadratic residue of m if

$$x^2 \equiv a \pmod{m}$$

has a solution x which is prime to m. If $(a, m) = 1$, and a is not a quadratic residue of m it is called a *quadratic nonresidue*. When the meaning is clear, we will sometimes merely say residue and nonresidue.

EXAMPLES: From

1	3	5	7
3	1	7	5
5	7	1	3
7	5	3	1

(mod 8)

1	3	7	9
3	9	1	7
7	1	9	3
9	7	3	1

(mod 10)

we see that 8 has 1 as its only quadratic residue, while 10 has both 1 and 9. From the table on page 61, 7 has 1, 2, and 4 as quadratic residues.

From Definition 12, page 33, it is clear that if $p \nmid a$, a is a quadratic residue of p, or is not, according as $\left(\frac{a}{p}\right) = +1$ or -1. Or, we may say, $\left(\frac{a}{p}\right) = +1$ or -1 according as a is or is not a square modulo p.

Theorem 30. *Every prime $p = 2P + 1$ has exactly P quadratic residues, and therefore also, P quadratic nonresidues.*

PROOF. In the proof of Euler's Criterion on page 36 we showed that if $\left(\frac{a}{p}\right) = +1$ there are exactly two incongruent solutions of $x^2 \equiv a \pmod{p}$.

Since each of the $2P$ classes $1, 2, \cdots, 2P$ has a square, there are exactly P distinct squares.

Definition 21. If $\left(\frac{a}{p}\right) = +1$ we write $\sqrt{a}$ (mod p) for either solution of $x^2 \equiv a \pmod{p}$. For $a = 0$, $\sqrt{0} = 0$. For $\left(\frac{a}{p}\right) = -1$, $\sqrt{a}$ does not exist modulo p.

EXERCISE 41. For every modulus m, the product of two residues is a residue, and the product of a residue and a nonresidue is a nonresidue. For every prime m and for *some* composite m, the product of two nonresidues is a residue, while for other composite m, the product of two nonresidues may be a nonresidue.

EXERCISE 42. Theorem 30 may be generalized to read that the number of residues is $\frac{1}{2}\phi(m)$ for some composite m, but not for others.

EXERCISE 43. For which primes $p = 24k + b$ does $\mathfrak{M}_p$ contain $\sqrt{-1}$, $\sqrt{2}$, or $\sqrt{3}$? Examine all eight possible combinations of the existence and the nonexistence of these square roots.

25. Is the Quadratic Reciprocity Law a Deep Theorem?

We interrupt the main argument to discuss a question raised on page 46. The *Quadratic Reciprocity Law* states that for any two distinct primes, $p = 2P + 1$ and $q = 2Q + 1$, p and q are both quadratic residues of each other, or neither is, unless PQ is odd. In that case, exactly one of the primes is a quadratic residue of the other. The theorem follows at once from Theorem 27 with the use of Definition 20 and Euler's Criterion.

The Quadratic Reciprocity Law is often refered to as a "deep" theorem. We confess that although this term "deep theorem" is much used in books on number theory, we have never seen an exact definition. In a qualitative way we think of a deep theorem as one whose proof requires a great deal of work—it may be long, or complicated, or difficult, or it may appear to involve branches of mathematics the relevance of which is not at all apparent. When the Reciprocity Law was first discovered, it would have been accurate to call it a deep theorem. But is it still?

Legendre's Reciprocity Law (so named by him), involves neither the concept of quadratic residues, nor the use of Euler's Criterion, as we have seen. With the simple proof given on page 45, we would not consider it a deep theorem.

Now divisibility questions of the form

$$q | N^2 - a$$

are clearly somewhat more involved than those of the form

$$q|a^Q - 1,$$

since $a^Q - 1$ is a *specific* number, while in $N^2 - a$, N is *unspecified* and may range over $2Q$ possibilities. Therefore it is not surprising that the Quadratic Reciprocity Law lies a *little* deeper than does Legendre's Reciprocity Law.

But even in the best of Gauss's many proofs, the theorem still seemed far from simple. It is of some interest to analyze the reasons for this.

(a) In his simplest proof, the third, Gauss starts with the "Gauss Lemma," (Exercise 30). From this, and a page or so of computation, he derives another formula. If a is odd:

$$\left(\frac{a}{p}\right) = (-1)^M \qquad (p = 2P + 1)$$

where

$$M = \sum_{x=1}^{P} \left[\frac{ax}{p}\right]. \tag{100}$$

Here [ ] is the greatest integer function, defined on page 14. Now it appears that with Eq. (100) Gauss has already dug deeper than need be. What we need is the parity of the *sum*, $\gamma + \gamma'$, (page 46). The *individual* exponent, M, is not needed, and, if it is obtained nonetheless, it is clear that this is not without some extra effort.

(b) Gauss then proceeds to prove that

$$\sum_{x=1}^{P} \left[\frac{qx}{p}\right] + \sum_{x=1}^{Q} \left[\frac{px}{q}\right] = PQ \tag{101}$$

by the use of various properties of the [ ] function. Here we see irrelevancies. What has the [ ] function to do with the Quadratic Reciprocity Law? Later Eisenstein simplified the proof of Eq. (101), but only by bringing in still another foreign concept—that of a geometric lattice of points. This is all very nice theory—but it all takes time.

(c) Finally there is a point which we may call "abuse of the congruence symbol." We have shown many uses of the notation, $\equiv \pmod{p}$. But this symbol may also be misused. Suppose we write Eq. (78) as follows:

$$qa \equiv r \pmod{p}, \tag{102}$$

and inquire as to the number of odd a's for which r is even. There are three things wrong with such an approach.

(1) We are interested not in one group $\mathfrak{M}_p$, but in the interrelation between two groups $\mathfrak{M}_p$ and $\mathfrak{M}_q$, and, for this, the congruence notation is not helpful.

(2) There are no "even" and "odd" residue classes. If a is even, then $a + p \equiv a$ is odd.

(3) Most important is the following. The concept "congruent to" is of value when, (as on page 54), we don't *care* what the quotient is. But in Eq. (78),

$$qa = pa' + r,$$

the *quotient* a', for the divisor p, is also a *coefficient* of p in evaluating $(p|q)$. And the quotient a is a coefficient of q for $(q|p)$. This is precisely where the *reciprocity* lies, and, if we throw it away, as in Eq. (102), we must work the harder to recover it.

Exercise 44. Evaluate $(13|17)$ by Eq. (100). Compare page 44.

26. Congruential Equations with a Prime Modulus

In Sects. 23 and 24 we developed reciprocals and square roots modulo m. With these we may easily solve the general linear and quadratic congruential equations for a prime modulus. These are

$$ax + b \equiv 0 \pmod p \qquad (p \nmid a) \tag{103}$$

and

$$ax^2 + bx + c \equiv 0 \pmod p \quad (p \nmid a). \tag{104}$$

The reader may easily verify that the solutions are the same as those given in ordinary algebra, that is,

$$x \equiv -a^{-1}b \qquad \pmod p \tag{103a}$$

and

$$x \equiv (2a)^{-1}(-b \pm \sqrt{b^2 - 4ac}) \qquad \pmod p. \tag{104a}$$

Therefore, "as" in ordinary algebra, Eq. (103) has precisely one solution, while Eq. (104) has 2, 1, or 0 solutions depending on whether $(b^2 - 4ac|p) = +1, 0, \text{ or } -1$.

Examples:

(a) $3x + 2 \equiv 0 \pmod 7$.
Since $3^{-1} \equiv 5$, $x \equiv -10 \equiv 4 \pmod 7$.

(b) $3x^2 + 4x + 1 \equiv 0 \pmod 7$.
Since $b^2 - 4ac = 4$ is a quadratic residue of 7, with square roots 2 and 5, we have $x \equiv 6^{-1}(-4 \pm 2) \equiv 6 \text{ or } 2 \pmod 7$.

(c) $2x^2 + 3x + 2 \equiv 0 \pmod 7$.
Since $9 - 16 \equiv 0 \pmod 7$ there is only one solution, namely $1 \pmod 7$.

The algebra here is so much like ordinary algebra because the residue classes modulo a *prime* form a *field*, just as the real or rational or complex numbers form a field. Thus, just as group theory applies to $\mathfrak{M}_m$ so does field theory apply here. An important theorem in field theory states that an nth degree polynomial can have at most n roots.

Theorem 31. *At most n residue classes satisfy the equation:*

$$f(x) = a_n x^n + a_{n-1}x^{n-1} + \cdots + a_0 \equiv 0 \pmod p \quad (105)$$

with $a_n \not\equiv 0 \pmod p$.*

Proof. Let Eq. (105) have n roots, $x_1, x_2, \cdots, x_n$. Dividing $f(x)$ by $x - x_1$ we obtain $f(x) = f_1(x)(x - x_1) + c_1$. But since $p | f(x_1)$ we find $p | c_1$. Therefore

$$f(x) = f_1(x)(x - x_1) + kp.$$

Repeating this operation with $f_1(x)$, then $f_2(x)$, etc., we obtain

$$f(x) = a_n(x - x_1)(x - x_2) \cdots (x - x_n) + pg(x)$$

for some polynomial $g(x)$. Now if there were an $n + 1$st root x_{n+1}, not congruent to one of the others, we would have

$$0 \equiv f(x_{n+1}) \equiv a_n(x_{n+1} - x_1)(x_{n+1} - x_2) \cdots (x_{n+1} - x_n) \pmod p$$

Therefore, by Theorem 6, Corollary, $a_n \equiv 0 \pmod p$, contradicting the hypothesis.

We will use this theorem later when we investigate *primitive roots*. We could have used it earlier, together with Fermat's Theorem, to prove Euler's Criterion.

If $N^2 \equiv a \pmod q$, then $N^{2Q} \equiv a^Q \pmod q$ and, by Fermat's Theorem, $a^Q \equiv 1 \pmod q$. The converse is the more difficult. But from Theorem 30 there are Q quadratic residues. Therefore, from what we have just shown, there are Q solutions of $a^Q - 1 \equiv 0 \pmod q$. But by Theorem 31, there can be no other solutions. Therefore $a^Q \equiv 1 \pmod q$ implies $N^2 \equiv a \pmod q$.

If p is *not* a prime, in Theorem 31, there may be a greater number of solutions. (Where does the proof break down?) Thus

$$x^2 \equiv 1 \pmod{24}$$

has 8 solutions, and so does

$$x^2 \equiv x \pmod{30}.$$

The equation $x^2 \equiv x \pmod m$ is particularly interesting, because in any

* Since $x^p \equiv x$, $x^{p+1} \equiv x^2$, etc., for every $x \pmod p$, any polynomial of order higher than $p - 1$ may be reduced to one of order not higher than $p - 1$.

field the two roots, 0 and 1, are the *identities* for addition and multiplication respectively. If m is divisible by more than one prime, we shall see that

$$x^2 \equiv x \pmod{m} \tag{106}$$

has more than 2 solutions, and that each one may serve as an identity element in a multiplicative group. Thus

$$x^2 \equiv x \pmod{10}$$

has 4 solutions 0, 1, 5, and 6. In addition to the set $\mathfrak{M}_{10}$, of elements 1, 3, 7, and 9, which form a group with 1 as the identity, so likewise 2, 4, 6, and 8 form a group modulo 10 with 6 as the identity, and 0 and 5 form groups of one element each, by themselves.

Exercise 45. Show that

$$2x^2 + 5x + 5 \equiv 0 \pmod{7}$$

has no solution.

Exercise 46. Find the 8 solutions of

$$x^2 \equiv x \pmod{30}$$

and show that corresponding to each solution there is a multiplicative group of residue classes, modulo 30, with that solution as the identity.

Exercise 47. Just as in Exercise 40, Eq. (99), we have an explicit formula for a reciprocal, a^{-1} modulo m, so, for some prime moduli, we have an explicit formula for a square root. Show that if $p = 4m + 3$, and $(a|p) = +1$, then $\sqrt{a} \equiv a^{m+1} \pmod{p}$. In particular

$$4^{k+1} \equiv \sqrt{2} \pmod{8k + 7}.$$

Also show

$$a^k + a^{-k} \equiv \sqrt{2} \pmod{8k + 1}$$

where a is any quadratic nonresidue of the prime $8k + 1$. Thus we may compute $\sqrt{2}$ explicitly for all the prime moduli for which it exists.

27. Euler's ϕ Function

On page 62 we noted that while certain theorems for $\mathfrak{M}_m$, with m a prime, could be extended to all $\mathfrak{M}_m$, or even to all finite groups; others, such as Euler's Criterion, could be extended to $\mathfrak{M}_m$ for *some* composite m, (say $m = 9, 10$), but not for others, (say $m = 8, 12$). In Exercises 41 and 42 there were closely related extensions, again valid only for some composite m. Likewise, back in Exercise 25 there was such an extension. We are concerned now with the underlying structural reasons for these differences. For this analysis we will want a better knowledge of Euler's ϕ function.

Our first result is

Theorem 32 (Euler). *If*

$$N = p_1^{a_1} p_2^{a_2} \cdots p_n^{a_n}, \tag{107}$$

then

$$\phi(N) = N\left(1 - \frac{1}{p_1}\right)\left(1 - \frac{1}{p_2}\right) \cdots \left(1 - \frac{1}{p_n}\right). \tag{108}$$

In the proof of Eq. (108) the main work is done (and constructively) by

Theorem 33. *If $A > 0$, $B > 0$, and $(A, B) = 1$, the AB numbers*

$$m = Ab + Ba$$

with $a = 0, 1, \cdots, A - 1$, and $b = 0, 1, \cdots, B - 1$, belong to distinct residue classes modulo AB. Further, if in m, the a's are confined to the $\phi(A)$ numbers prime to A, and the b's to the $\phi(B)$ numbers prime to B, then the corresponding $\phi(A)\phi(B)$ numbers m are all prime to AB.

Proof. If

$$Ab_1 + Ba_1 \equiv Ab_2 + Ba_2 \pmod{AB}$$

then $Ab_1 \equiv Ab_2 \pmod{B}$ and $Ba_1 \equiv Ba_2 \pmod{A}$. But since $(A, B) = 1$' by Theorem 29, $b_1 \equiv b_2 \pmod{B}$ and $a_1 \equiv a_2 \pmod{A}$. Furthermore'

$$\mu = A\beta + B\alpha \equiv A\beta \pmod{B}.$$

Since $(A, B) = 1$, and if $(\beta, B) = 1$, we have μ prime to B. Likewise if $(\alpha, A) = 1$, μ is prime to A. Therefore if $(\alpha, A) = (\beta, B) = 1$, μ is prime to AB.

Corollary. *If $A > 0$, $B > 0$, and $(A, B) = 1$, then*

$$\phi(AB) = \phi(A)\phi(B). \tag{109}$$

Proof. The $\phi(A)\phi(B)$ numbers μ just indicated are prime to AB, and not congruent modulo AB. Furthermore, each such μ is congruent to exactly one integer satisfying $0 < x < AB$. No other of the AB numbers $m = Ab + Ba$ are prime to AB, for if $(a, A) \neq 1$, then m is not prime to A, nor therefore to AB. Similarly for b and B. This proves Eq. (109).

Proof of Theorem 32. If $N_1 = p_1^{a_1}$, the numbers $\leqq N_1$ and not prime to N_1 are the multiples of $p_1 \leqq N_1$. Since there are $p_1^{a_1 - 1}$ of these, we have

$$\phi(N_1) = N_1\left(1 - \frac{1}{p_1}\right),$$

and by applying Eq. (109) $n - 1$ times we obtain Eq. (108).

Another important result concerning $\phi(N)$ is

Theorem 34 (Gauss). *If* $N > 0$,

$$\sum_d \phi(d) = N, \tag{110}$$

where the sum on the left is taken over all positive divisors d *of* N.

EXAMPLE: $N = 341$ has four positive divisors, 1, 11, 31, and 341.

$$\phi(1) + \phi(11) + \phi(31) + \phi(341) = 1 + 10 + 30 + 300 = N.$$

PROOF. Consider the equation

$$(x, N) = \frac{N}{d} \tag{111}$$

where d is a positive divisor of N and x can be $1, 2, \cdots, N$. Any solution x of Eq. (111) must be a multiple of $\frac{N}{d}$, $x = k\frac{N}{d}$, where $1 \leqq k \leqq d$. Further any such x will be a solution if and only if $(k, d) = 1$, since $mk + nd = 1$ implies $m\left(k\frac{N}{d}\right) + nN = \frac{N}{d}$ and conversely. There are therefore $\phi(d)$ solutions. Since every $1 \leqq x \leqq N$ satisfies an Eq. (111) for one and only one d, we obtain Eq. (110).

Theorems 32 and 34 could lead us off in several directions. Thus

(a) From Theorem 32, Euler proved Euclid's Theorem 8 as follows. If

$$M = 2 \cdot 3 \cdot 5 \cdots p_n$$

and if there were no primes $<M$ other than $2, 3, \cdots, p_n$ we would have $\phi(M) = 1$. But $\phi(M) = M\left(1 - \frac{1}{2}\right)\left(1 - \frac{1}{3}\right) \cdots \left(1 - \frac{1}{p_n}\right) > 1$.

On the other hand, we *now* have an upper bound:

$$\pi(M) \leqq \phi(M) + n.$$

As $n \rightarrow \infty$, we see that

$$\frac{\phi(M)}{M} = \left(1 - \frac{1}{2}\right)\left(1 - \frac{1}{3}\right) \cdots \left(1 - \frac{1}{p_n}\right) \tag{112}$$

decreases monotonically and, if we are investigating $\pi(N)$ we are led to the question of estimating the right side of Eq. (112).

(b) Perhaps it was such a consideration which led Euler to his famous identity:

$$\sum_{n=1}^{\infty} \frac{1}{n^s} = \prod_p \left(1 - \frac{1}{p^s}\right)^{-1} \quad (s > 1) \tag{113}$$

wherein the infinite product on the right is taken over all primes. This identity, in the hands of Riemann and others, led eventually to a proof of Theorem 9. If $s = 1$, the harmonic series on the left, $1 + \frac{1}{2} + \frac{1}{3} + \cdots$, diverges. If there were only a finite number of primes, the product on the right would remain finite and yet equal to the series on the left. This contradiction gives another proof of Theorem 8. Again, if $s = 2$, we have

$$\sum_{n=1}^{\infty} \frac{1}{n^2} = \frac{\pi^2}{6} = \prod_p \left(1 - \frac{1}{p^2}\right)^{-1},$$

so that if Theorem 8 were false we would have π^2 equal to a rational number. This is known to be false, and if this latter does not *already* assume Theorem 8, we have still another proof.

(c) Equation (108) also leads to mean value theorems for $\phi(N)$ and $\phi(N)/N$ as $N \to \infty$, and to an interesting relationship between $\phi(N)$ and $\sigma(N)$, the sum of the positive divisors of N.

(d) Theorems 32 and 34 have a relationship, via the so-called *Möbius Inversion Formula*, which has an important generalization.

But we shall follow none of these diverging leads at this time. What is now in order is a deduction of *primitive roots* using Theorem 34.

EXERCISE 48. Since $(a, N) = 1$ implies $(N - a, N) = 1$, $\phi(N)$ is even if $N > 2$.

EXERCISE 49. Verify Theorem 34 for $N = 561$.

EXERCISE 50. Verify Theorem 34 for $N = 30$. What is the relationship between this partition of 30 and that of Exercise 46? HINT: Compare the proof of Theorem 34 with the membership in the eight groups.

EXERCISE 51. Find several multiplicative groups modulo 561 other than $\mathfrak{M}_{561}$.

28. PRIMITIVE ROOTS WITH A PRIME MODULUS

For every a prime to m

$$a^{\phi(m)} \equiv 1 \pmod{m},$$

but for some a, a smaller exponent, s, may suffice for

$$a^s \equiv 1 \pmod{m}$$

to be satisfied. Thus for any quadratic residue of m, (if $m > 2$) we have

$$a^{\phi(m)/2} \equiv 1 \pmod{m}.$$

Definition 22. If $(a, m) = 1$ and e is the *smallest* positive exponent such that

$$a^e \equiv 1 \pmod{m} \tag{114}$$

we say a is *of order* e modulo m.

Example: If $a = 10$ and m is a prime $\neq 2$ or 5, then the order e is also the *period* of the periodic decimal $1/m$. Thus 10 is of order 3 modulo 37, as on page 55. (It is probable that this definition, and Definition 23, Theorem 35, and Theorem 36 which follow, all stem from Gauss's early studies in periodic decimals mentioned on page 53. See Exercise 8S on page 203 for a plausible reconstruction of Gauss's line of thought.)

Theorem 35. *If $(a, m) = 1$, and a is of order e, then if*

$$a^f \equiv 1 \pmod{m}$$

we have $e|f$. In particular $e|\phi(m)$. Further, $a^1, a^2, a^3, \cdots, a^e$ belong to e distinct residue classes modulo m.

Proof. We have $m|a^e - 1$ and $m|a^f - 1$, and by Theorem 10, $m|a^g - 1$ where $g = (e, f)$. Therefore $g \leqq e$. But $g \nless e$ by the definition of e. Therefore $e = g$ and $e|f$. Further, if $a^{e_1} \equiv a^{e_2} \pmod{m}$, and $e \geqq e_1 > e_2 \geqq 1$, we have $a^{e_1 - e_2} \equiv 1 \pmod{m}$, which again contradicts the definition of e.

Theorem 36 (Gauss). *If $d|p - 1$, where p is a prime, there are $\phi(d)$ residue classes of order d modulo p.*

Proof. From Theorem 35, if a is of order e modulo p, then $a^1, a^2, a^3, \cdots, a^e$ are e distinct residue classes. They are thus e distinct solutions of

$$x^e \equiv 1 \pmod{p},$$

and, by Theorem 31, there can be no others. Each class of order e modulo p is therefore contained among these e classes. But if $r \leqq e$ and $(r, e) \neq 1$, let $r = sg$ and $e = tg$ with $g > 1$. Then

$$(a^r)^t = (a^e)^s \equiv 1 \pmod{p},$$

and we find that a^r is of order $\leqq t < e$. Let $\psi(e)$ be the actual number of classes of order e. Then, by Theorem 35, if $e \nmid p - 1$, $\psi(e) = 0$, and if $e|p - 1$, we have just shown that

$$\psi(e) \leqq \phi(e). \tag{115}$$

But since every class, $1, 2, \cdots, p - 1$ is of some order modulo p we have

$$\sum_d \psi(d) = p - 1$$

where the sum is taken over all positive divisors of $p - 1$. Since from Theorem 34 we now have

$$\sum_d [\phi(d) - \psi(d)] = 0,$$

and since, from Eq. (115), each $[\phi(d) - \psi(d)] \geqq 0$, we obtain

$$\psi(d) = \phi(d)$$

for every d.

Definition 23. If $(a, m) = 1$, and a is of order $\phi(m)$ modulo m, we call a a *primitive root* of m. In particular, for a prime modulus p, a primitive root of p is a residue class of order $p - 1$.

EXAMPLE: Since, on page 53, the decimal expansion of $\frac{1}{7}$ is of period 6, $3 \equiv 10$ is a primitive root of 7.

The importance of Theorem 36 is that it guarantees (nonconstructively!) a primitive root for every prime modulus. This result—that is, every prime modulus has a primitive root—is one of the fundamental theorems of number theory. It is the basis of the theorems which we shall obtain in this chapter concerning the structure of the $\mathfrak{M}_m$ groups. In particular, it is the basis of the structural differences which we sought at the end of Sect. 23 and the beginning of Sect. 27. It implies that $\mathfrak{M}_p$ is a *cyclic* group.

EXERCISE 52. For every divisor d of 12, determine the $\phi(d)$ residue classes of order d modulo 13, in particular, determine the 4 primitive roots of 13.

EXERCISE 53. For every prime $p > 2$, 1 is of order 1 and $p - 1$ is of order 2 and these are the sole residue classes of these orders.

29. $\mathfrak{M}_p$ AS A CYCLIC GROUP

Definition 24. A group is *cyclic* if it contains an element g, called a *generator*, such that every element a in the group may be expressed as

$$a = g^n$$

for some integral exponent, positive, negative, or zero.

By Theorem 36, p has $\phi(p - 1)$ distinct primitive roots. Let g be any one of these. Since, by the last sentence of Theorem 35, $g, g^2, \cdots, g^{p-1}$ are all distinct, g serves as a generator for $\mathfrak{M}_p$, and thus $\mathfrak{M}_p$ is cyclic. By rearranging the rows and columns of the table for $\mathfrak{M}_7$ on page 61, and since 3 is a primitive root of 7, we obtain

1	3	2	6	4	5
3	2	6	4	5	1
2	6	4	5	1	3
6	4	5	1	3	2
4	5	1	3	2	6
5	1	3	2	6	4

where the kth element in the first row is congruent to 3^{k-1}. Here 3 is the generator, and the $(n + 1)$st row is obtained from the first by a left, n shift, cyclic permutation.

Some composite m may also have a primitive root; thus 2 is one for 9.

1	2	4	8	7	5
2	4	8	7	5	1
4	8	7	5	1	2
8	7	5	1	2	4
7	5	1	2	4	8
5	1	2	4	8	7

(mod 9)

For *any* modulus $m > 2$ which possesses a primitive root g, regardless of whether m is prime or composite, it is almost immediate that if $a \equiv g^n \pmod{m}$, then a is a quadratic residue of m or not according as n is even or odd. Further, there are exactly $\frac{1}{2}\phi(m)$ residues. Further (Euler's Criterion generalized), a is a residue if and only if $a^{\frac{1}{2}\phi(m)} \equiv 1 \pmod{m}$. Further, the product of two nonresidues is a residue. We will determine later which composite m have a primitive root, and therefore also these other properties.

EXERCISE 54. Prove the "if" part of Wilson's theorem (page 37) using a primitive root of the prime q. HINT: evaluate the sum $1 + 2 + \cdots + (q - 1)$ modulo $q - 1$. With reference to Exercise 25, generalize the proof here to those composites m which have primitive roots.

Definition 25. Two groups $\mathcal{A}$ and $\mathcal{B}$ are said to be *isomorphic* if every element a of $\mathcal{A}$ may be put into one-to-one correspondence with an element b of $\mathcal{B}$,

$$a \leftrightarrow b$$

in such a way that if $a_1 \leftrightarrow b_1$ and $a_2 \leftrightarrow b_2$, then $a_1a_2 \leftrightarrow b_1b_2$. That is, the correspondence is preserved under the group operation. Starting with the a's and performing first the mapping, $a \rightarrow b$, and then the product, we will obtain the same result as if we first perform the product, and then the mapping. In an isomorphism, therefore, these two operations may be commuted. If $\mathcal{A}$ and $\mathcal{B}$ are isomorphic, we write

$$\mathcal{A} \rightleftharpoons \mathcal{B}$$

and we consider them to be the *same* "abstract group."

It is easily seen that two *cyclic* groups of the same order are always isomorphic. Thus

$$\mathcal{M}_7 \rightleftharpoons \mathcal{M}_9$$

under the mapping

$$\pmod{7} \qquad 3^n \leftrightarrow 2^n \qquad \pmod{9}.$$

Or, if we prefer, under

$$(\text{mod } 7) \qquad 5^n \leftrightarrow 5^n \qquad (\text{mod } 9),$$

since 5 is a primitive root of both moduli.

The group of the mth roots of unity, $e^{2\pi ai/m}$, for $a = 0, 1, \cdots, m-1$, under ordinary multiplication; the group of rotations of the plane through $360a/m$ degrees, for $a = 0, 1, \cdots, m-1$, under addition of angles; and the group of the m residue classes under *addition* modulo m, are all isomorphic. They all are the same abstract group—namely, the cyclic group of order m. We designate this group as $\mathcal{C}_m$.

The isomorphism between $\mathfrak{M}_p$ for a prime p and $\mathcal{C}_{p-1}$ suggests a circular representation of $\mathfrak{M}_p$, which eliminates the obvious redundancy in the multiplication table for $\mathfrak{M}_p$, and which we illustrate for $p = 17$:

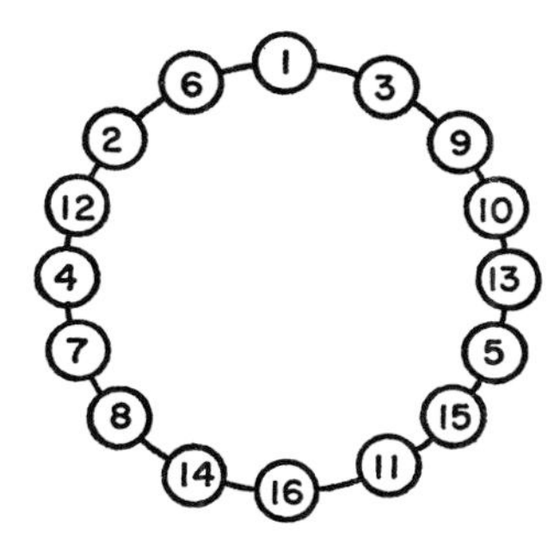

Here 3 is the generator and successive powers of 3 correspond to successive rotations thru $22\frac{1}{2}°$. Or $3^{-1} \equiv 6$ may be considered the generator and its powers are strung out in the opposite direction. Two residue classes at angles α and β have a product at an angle $\alpha + \beta$. In particular, reciprocals lie at an equal distance from 1 in opposite directions. The residue $-1 \equiv 16$ is thus its own reciprocal, and the only class of order 2. It follows that residues on opposite ends of a diameter add to 17; each is congruent to the other's negative. The quadratic residues are 1, 9, 13, 15, etc.

It is well known that historically $\pm i = \sqrt{-1}$ did not attain full respectability until it was interpreted as a rotation of 90°. If p is an odd prime, $\mathfrak{M}_p$ will have a $\sqrt{-1}$ if and only if $p = 4m + 1$. We now see the significance of this, in that only $\mathcal{C}_{4m}$ allows a rotation of exactly 90°. Thus for $p = 17$, in the diagram, we have 4 and 13 as the two values of $\sqrt{-1}$.

We see also that Euler's Criterion,

$$a^{P} \equiv \left(\frac{a}{p}\right) (\text{mod } p),$$

and his even more celebrated formula,

$$e^{n\pi i} = (-1)^n,$$

are very intimately related. Euler was no doubt the world's most prolific mathematician. A modern mathematician, looking at the last two equations, may be tempted to say, "No wonder, he works both sides of an isomorphism." But better judgment at once prevails—had Euler not worked both sides, the isomorphism may not have been discovered.

EXERCISE 55. Show that $\mathfrak{M}_{14} \rightleftharpoons \mathfrak{M}_9$ and give two distinct mappings.

EXERCISE 56. Show that other circular representations of $\mathfrak{M}_{17}$ may be obtained from the given one by starting at 1 and taking steps of $k \cdot 22\frac{1}{2}°$ where $(k, 16) = 1$. More generally, if g is a primitive root of p, g^k is also, if and only if $(k, p - 1) = 1$.

EXERCISE 57. Show that

$$\mathfrak{M}_8 \rightleftharpoons \mathfrak{M}_{12}$$

but

$$\mathfrak{M}_8 \not\rightleftharpoons \mathfrak{M}_{10}.$$

Show that $\mathfrak{M}_8$ is not cyclic.

30. THE CIRCULAR PARITY SWITCH

In 1956 the author invented the following unusual switch.

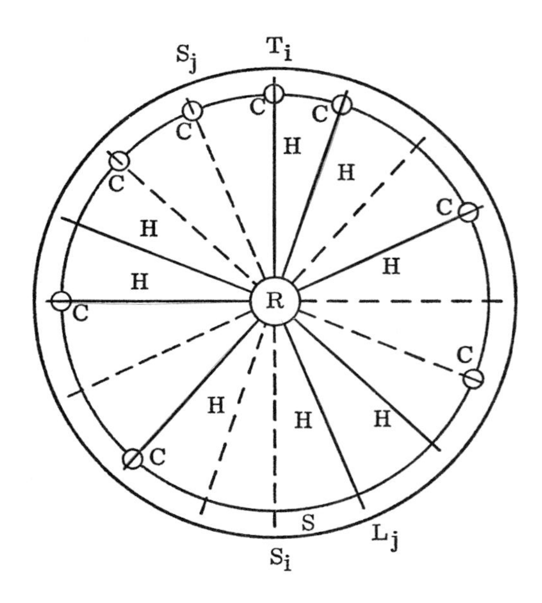

Definition 26. A *circular parity switch* of order N has a stator (S) with $2N$ equally spaced divisions. At N of these there are contacts (C). Their

locations are arbitrary except that no two contacts lie on a diameter. There is a rotor (R) which may assume the $2N$ angular positions, and attached rigidly to R, at any of the $2N$ divisions on the hub, are N hands (H). Again, their location is arbitrary except that no two lie in the same diameter. Let m hands be touching contacts in a particular position of the rotor.

Theorem 37. *As the rotor turns, (in either direction), m will be alternately even and odd.*

EXAMPLE: In the special case for $N = 8$ in the diagram, a clockwise rotation will give the following periodic m sequence: 5, 2, 5, 4, 5, 4, 3, 4, 3, 6, 3, 4, 3, 4, 5, 4, repeat.

PROOF. Opposite each hand in a rotor is a *space*. Let a complete group of contiguous hands with no spaces in between be called a *bunch*, and reading clockwise let the first hand in a bunch be called a *trailing hand*, and the last hand, a *leading hand*. Let a complete group of contiguous spaces be called a *gap*. Put each trailing hand T_i into correspondence with the leading hand L_j immediately preceding the space opposite T_i. There is such an L_j since preceding T_i there is a space S_j. Opposite S_j is a hand. Since this is followed by the space S_i which is opposite T_i, the hand is a leading hand.

Now as the rotor turns one division (clockwise), the only changes in m which need be counted are those in which a leading hand picks up a contact or a trailing hand drops one. For if a nonleading hand picks up a contact, it was dropped by the hand ahead of it; and if a nontrailing hand drops a contact, it is picked up by the hand behind it. But there was either a contact under T_i or in S_i, but not both. Therefore either T_i will drop this contact, or L_j will pick it up, but not both. The contribution of the pair of hands towards changing m is therefore ± 1.

But starting at T_i, and going clockwise to L_j, we will pass k bunches and $k - 1$ gaps. And the remaining bunches in the other half of the rotor may be reflected into these $k - 1$ gaps. Thus the total number of bunches, $2k - 1$, is odd, and the number of pairs, T_i and L_j, is therefore also odd. But a change in m by an odd number of ± 1 means a change of parity.

We now ask, how many distinct rotors of order N are there—that is, rotors that cannot be transformed into each other merely by rotation? Call this number $R(N)$. If N is an odd prime, we obtain an old friend.

Set aside the special rotor R_1 consisting alternately of one hand and one space. Consider any other rotor of order N, and in particular consider the pattern of hands and spaces in a block of N consecutive divisions. This pattern may be represented by an N-bit binary number, with ones for hands, and zeros for spaces. Excluding the two possible patterns in R_1:

$$1010 \cdots 01$$

and $$0101 \cdots 10,$$

any of the $2^N - 2$ remaining patterns is a legitimate one, and will occur in precisely one rotor R_i. It cannot occur in two, since the remaining N divisions of R_i must have the complementary pattern, and therefore R_i is completely defined. If a different block of N consecutive divisions in R_i is examined, a different pattern must be found. For if two patterns in R_i were identical, R_i would have to be periodic, with a period less than $2N$. This period must divide $2N$. The period cannot be the prime N, since we know that complementary blocks of N divisions have complementary patterns, not the same. The period cannot be 2, since we excluded those two patterns. Thus R_i must have $2N$ different patterns. Therefore

$$R(N) = \frac{2^N - 2}{2N} + 1, \tag{116}$$

and since $R(N)$ is an integer, we have reproven Fermat's Theorem 12.

A second application of the parity switch is this. Consider the circular diagram for $\mathfrak{M}_p$ (page 75) as a stator, with contacts at the even numbers. This is a legitimate stator since opposite each even e is the odd $p - e$ as we showed on page 75. Let the rotor have hands which, in one position, point to every odd number. If the hand pointing to 1 is now brought around to the number a, the $(p - 1)/2$ hands will point to the $(p - 1)/2$ products

$$1 \cdot a,\ 3 \cdot a,\ 5 \cdot a,\ \cdots,\ (p - 2) \cdot a \qquad (\text{mod } p),$$

and let m of these products be even. Since in the rotor's original position m is 0, by Theorem 37 m will be even or odd according as a is a quadratic residue or not. That is,

$$\left(\frac{a}{p}\right) = (-1)^m.$$

Thus we have reproven a combination of Euler's and Gauss's Criteria with the aid of a switch.

31. Primitive Roots and Fermat Numbers

By characterizing $\mathfrak{M}_p$ as a cyclic group, for every prime p, we have gone the limit in its *structural* analysis. A cyclic group is the simplest type; and we may say that there remain no questions concerning its structure. But the content of that structure is quite another matter. Thus we know, at once, that $\mathfrak{M}_7 \rightleftarrows \mathfrak{C}_6$:

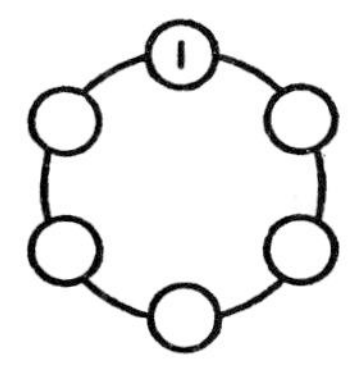

But until we compute a primitive root we cannot (completely) assign the residue classes to suitable billets. (Where $p - 1 \equiv -1$ goes is simple enough.)

Given a prime p, it is always possible to compute a primitive root by trial and error, since $\mathfrak{M}_p$ is finite. For $p > 2$, a quadratic residue of p is clearly not a primitive root of p. For if a is a quadratic residue of $p = 2P + 1$ we have $a^P \equiv 1 \pmod{p}$ by Euler's Criterion. Thus the order of a modulo p is P or smaller. Further, for $p > 3$, $p - 1 \equiv -1$ is not a primitive root, since $(-1)^2 \equiv 1$. But with these obvious exceptions, and with no deeper theory, one might now examine the remaining residue classes in search of a primitive root. Gauss, and others, have devised more efficient techniques, but no general, *explicit*, *nontentative* method has been devised, and this, like a good criterion for primality, remains an important unsolved problem.

The converse problem is even harder. Given an integer g, for which primes p is g a primitive root? Not even in a single instance is it known that there are infinitely many such primes p. For example consider

Theorem 38. *If $p = 4m + 3$ and $q = 2p + 1$ are both prime, -2 is a primitive root of q.*

EXAMPLE:

$$-2 \equiv 5 \text{ is a primitive root of } 7.$$

PROOF. There are $\phi(2p) = p - 1$ primitive roots of q. None of the p quadratic residues, a, of q can be a primitive root, as above. Nor can -1, which is not a quadratic residue, be a primitive root. Thus any *other* quadratic nonresidue is a primitive root, and -2 is always one, since

$$(-2|q) = -(2|q)_{MN} = -1_D.$$

Therefore if Conjecture 4 were true we could prove the existence of infinitely many q with -2 as a primitive root.

Similarly, if the weaker Conjecture 5 were true, we could utilize

Theorem 39. *If p and $q = 2p + 1$ are both odd primes, -4 is a primitive root of q.*

EXAMPLE:

$$-4 \equiv 3 \text{ is a primitive root of } 7.$$

The proof of Theorem 39 is left for the reader. Another theorem of slightly different character is

Theorem 40. *If* $F_m = 2^{2^m} + 1$ *is a prime, with* $m \geqq 1$, 3 *is a primitive root of* F_m.

EXAMPLE:

$$3 \text{ is a primitive root of } 5 = F_1 \text{ and of } 17 = F_2.$$

PROOF. Since $\phi(F_m - 1) = \frac{1}{2}(F_m - 1)$, we see that in this (unusual) case any quadratic nonresidue of F_m is also a primitive root. But

$$F_m \equiv 5 \pmod{12}$$

by induction, since $F_1 = 5$, and

$$F_{m+1} = (F_m - 1)^2 + 1.$$

Therefore, by Theorem 20, $(3|F_m) = -1$.

Here, again, we do not know whether there are infinitely many *Fermat numbers*, F_m, which are prime. Fermat thought all F_m might be prime, but said he couldn't prove it. Euler showed, however, that $641|F_5$, as on page 58. Aside from the five primes, F_m for $0 \leqq m \leqq 4$, no other prime F_m has been found. On the contrary, F_m for $5 \leqq m \leqq 16$, at least, are all composite. Any prime F_m corresponds to a constructable regular polygon, (Gauss, page 52). Like the Mersenne numbers, (page 18), the Fermat numbers, (page 13), are all prime to each other.

There are three possibilities:

(a) Only finitely many F_m are composite.

(b) Only finitely many F_m are prime.

(c) Infinitely many F_m are prime, and infinitely many are composite.

If (a) or (c) were true, we could find infinitely many primes with 3 as a primitive root, but actually possibility (b) is the most likely. We will return to this question in Exercise 36S, page 214.

EXERCISE 58. Criticize the word "explicitly" in the last sentence in Exercise 47. Investigate possibilities of remedying this flaw.

EXERCISE 59. Find a primitive root of $p = 41$.

EXERCISE 60. Find 16,188,302,110 primitive roots of $q = 32{,}376{,}604{,}223$.

EXERCISE 61. If $p = 4m + 3 > 3$ and $q = 2p + 1$ are both primes, there are at least three successive integers, g, $g + 1$, and $g + 2$, which are all primitive roots of q.

EXERCISE 62. Using residue arithmetic, show that

$$274177|F_6.$$

32. ARTIN'S CONJECTURES

It is easily seen that -1 is a primitive root only for the primes 2 and

3; +1, and all odd squares, are primitive roots only for the prime 2; and any even square is never a primitive root. In spite of the negative results of the previous section, the evidence is sufficient to warrant our stating

Conjecture 13 (Artin). *Every integer a, not equal to -1 or to a square, is a primitive root of infinitely many primes.*

It is likely that a stronger result is true:

Conjecture 14 (Artin). *If $a \neq b^n$ with $n > 1$, and if $\nu_a(N)$ is the number of primes $\leqq N$ for which a is a primitive root, then*

$$\nu_a(N) \sim 0.3739558\ \pi(N). \tag{117}$$

This conjecture was made by E. Artin in a conversation with H. Hasse in 1927. It states that for $a = 2, 3, 5, 6, 7, 10$, etc., approximately $\frac{3}{8}$ of all primes will have a as a primitive root, and that this asymptotic ratio, $0.37 \cdots$, is independent of a. (If a is a cube or some other odd power, there is a minor complication, which need not concern us here.)

We shall explain presently the coefficient in Eq. (117), and the heuristic reasoning behind Eq. (117). But first we examine two tables based on counts, $\nu_a(N)$, given by Cunningham (1913).

a	$\nu_a(10{,}000)$	$\nu_a(10{,}000)/\pi(10{,}000)$
2	470	.3824
3	476	.3873
5	492	.4003
6	470	.3824
7	465	.3784
10	467	.3800
11	443	.3605
12	459	.3735
		.3806 av.

$N/10{,}000$	$\nu_2(N)$	$\nu_2(N)/\pi(N)$	$\nu_{10}(N)$	$\nu_{10}(N)/\pi(N)$
1	470	.3824	467	.3800
2	840	.3714	865	.3824
3	1205	.3713	1234	.3803
4	1570	.3735	1587	.3776
5	1923	.3746	1947	.3793
6	2263	.3736	2296	.3791
7	2589	.3733	2639	.3805
8	2928	.3736	2975	.3796
9	3274	.3758	3291	.3777
10	3603	.3756	3618	.3772
		.3745 av.		.3794 av.

In the smaller table we see that ν_a is substantially independent of a for the eight smallest positive integers not equal to a power. In the larger table, for the two most studied cases, $a = 2$ (related to perfect numbers), and $a = 10$ (related to periodic decimals), we see that $\nu_a(N)/\pi(N)$ changes only slightly with N.

A probability argument which makes Conjecture 14 plausible runs as follows. Consider $a = 2$, and the primes $p \leqq N$. For every p choose a primitive root g and write $g^m \equiv 2 \pmod{p}$ and $(m, p - 1) = G$. What is the probability that $2|G$? Except for $p = 2$, $p - 1$ is always even, and m is even in one half the cases—that is, when 2 is a quadratic residue of p. Since G must be 1 if 2 is to be a primitive root of p, we delete these cases, leaving, in the mean, $(1 - \frac{1}{2})\pi(N)$ primes. What is the probability that $3|G$? Except for $p = 3$, all primes are $3k + 1$ or $3k + 2$, and therefore $3|p - 1$ in one-half the cases, while $3|m$ in one third the cases. Eliminating the remaining primes where $3|G$ we are left with $\left(1 - \frac{1}{2}\right)\left(1 - \frac{1}{3 \cdot 2}\right)\pi(N)$ primes. Continuing with $5|G$, $7|G$, etc., we are left with

$$A \cdot \pi(N)$$

primes with $G = 1$, where the coefficient A (called Artin's constant), is given by the infinite product:

$$A = \prod_p \left(1 - \frac{1}{p(p-1)}\right). \tag{118}$$

The argument may be improved somewhat by using Theorem 16 and analogous results, but this improvement does not suffice to constitute a real proof of Conjecture 14. For any other *nonpower* a, the argument is unchanged, but for $a = 8$, say, we have $3|m$ in all the cases where $p = 3k + 1$. This changes the factor $\left(1 - \frac{1}{3 \cdot 2}\right)$ to $\left(1 - \frac{1}{2}\right)$ and we find instead

$$\nu_8(N) \sim \tfrac{3}{5}A\pi(N), \text{ etc.}$$

J. W. Wrench, Jr., has recently completed a highly accurate computation of Artin's constant. He gets

$$A = 0.37395\ 58136\ 19202\ 28805\ 47280\ 54346\ 41641\ 51116 \cdots. \tag{119}$$

If Artin's Conjecture 14 proves as obdurate as the conjectures of Sect. 12—and there is little doubt that it will—Wrench's Eq. (119) should suffice as a check on any empirical studies of $\nu_a(N)$ for quite a long time.

There is distinct tendency for $\nu_a(N)/\pi(N)$ to run high for small values of N—that is, for this ratio to approach A from above, aside from fluctuations. This may be noted in both tables above, and also, more clearly, in the following data: $\nu_2(N)/\pi(N) = 0.3988, 0.3861, 0.3857$, and 0.3849 for $N = 1000, 2000, 3000$, and 4000. This tendency has an interesting explanation. If a prime does not have 2 as a primitive root, the reason, four times out of five, is that $(2|p) = +1$. These latter primes are those of the forms $8k \pm 1$. While it is true that these primes are equinumerous to those of the forms $8k \pm 3$, nonetheless there is a definite tendency for the class of primes $8k + 1$ to lag behind the other three classes. See page 21 for some data. This interesting lag (which we will discuss in Volume II) has the consequence that $(1 - \frac{1}{2})$, the first factor in A, is too small for these modest N, and therefore, in general, $\nu_2(N)$ runs too high.

33. Questions Concerning Cycle Graphs

We now concern ourselves with the structure of $\mathfrak{M}_m$ with m not necessarily a prime. A good insight into these structures will be gained by the study of the *cycle graphs* of these groups.

Definition 27. If $(a, m) = 1$ and a is of order e modulo m, the e residue classes $a^1, a^2, a^3, \cdots, a^e$ are called the *cycle* of a modulo m. The definition may be clearly generalized to any finite group.

Definition 28. If a set $\mathcal{S}$ of elements in a group $\mathcal{G}$ is closed under the group operation, and contains the identity and the inverse of each of its elements, it is called a *subgroup* of $\mathcal{G}$. In particular, $\mathcal{G}$ itself is also a subgroup of $\mathcal{G}$.

It is clear that each cycle of $\mathfrak{M}_m$ is a cyclic subgroup of $\mathfrak{M}_m$. A diagram of a group, which shows every cycle in the group, and the *connectivity* among these cycles, is called a *cycle graph* of the group. It generalizes the circular diagram of $\mathfrak{M}_{17}$ on page 75. On pages 87–92 we show cycle graphs for 14 nonisomorphic $\mathfrak{M}_m$ groups. We will first make some comments, and we will then raise some questions.

Let our point of departure be the cycle graph of $\mathfrak{M}_{55}$ on page 88. It is of only moderate complexity, and thus is best adapted to illustrate the concept. The powers of 2 (mod 55), namely 1, 2, 4, 8, 16, 32, 9, 18, etc., constitute the cycle of 2 modulo 55. This cyclic subgroup of $\mathfrak{M}_{55}$ is of order 20, and is easily seen in the graph. Now $53 \equiv -2 \pmod{55}$ is not in this subgroup. Therefore the cycle of 53, which is also of order 20, is connected to the cycle of 2 only at their even powers, that is, at the quadratic residues. Similarly $51 \equiv -4$ has a cycle of order 10 which is connected to that of 4 at *their* even powers. Finally, the cycle of 29 completes the $40 = \phi(55)$ residue classes in $\mathfrak{M}_{55}$. No residue class is of order 40 modulo 55 and therefore $\mathfrak{M}_{55}$ is not cyclic.

Now let us back up to some smaller composite moduli. The smallest m

for which $\mathfrak{M}_m$ is not cyclic is 8. This is a well-known group of order 4—the "Four" group. Here 3, 5, and 7 are all of order 2, and their 3 cycles are connected only at their common square, 1. Since $\mathfrak{M}_{12} \rightleftarrows \mathfrak{M}_8$, their cycle graphs look alike—in fact, if 3 is replaced by 11, they are identical.

The next noncyclic group is $\mathfrak{M}_{15}$. Here four residue classes are of the highest order, 4, and the cycles for 2 and 7, say, are connected at their common square, 4, and common fourth power, 1. Two other cycles are those of 11 and 14. It is clear that in the cycle graphs we are concerned only with the ordering in, and topology of, the cycles. The actual size, shape, or location of the various cycles is not meant to be of significance. As with the circular diagram for $\mathfrak{M}_{17}$, we can easily read off the powers, order, and inverse of every residue class.

It may be seen that

$$\mathfrak{M}_{15} \rightleftarrows \mathfrak{M}_{16} \rightleftarrows \mathfrak{M}_{20} \rightleftarrows \mathfrak{M}_{30}.$$

$\mathfrak{M}_{24}$ is also of order 8, but is not isomorphic to $\mathfrak{M}_{15}$, or to any other $\mathfrak{M}_m$. It has only one quadratic residue.

$\mathfrak{M}_{21}$, which is isomorphic to $\mathfrak{M}_{28}$, $\mathfrak{M}_{36}$, and $\mathfrak{M}_{42}$, may be generated by the three cycles of 10, 11, and 17. These three cycles are connected at the three quadratic residues.

$\mathfrak{M}_{54}$ is cyclic and isomorphic to $\mathfrak{M}_{19}$.

$\mathfrak{M}_{63}$ really needs three dimensions. The *four* bunches, of *three* cycles each, regroup, after passing through the quadratic residues 4, 25, 37, and 22, into *three* bunches of *four* cycles each. After passing through the square roots of unity 62, 8, and 55, they again regroup, etc. By "needs three dimensions" we mean, of course, that it cannot be drawn in two dimensions without some cycles crossing each other. In three dimensions $\mathfrak{M}_{63}$ may be neatly represented as four $\mathfrak{M}_{21}$-like structures, in four planes separated by angles of 45°, and joined together at the four square roots of unity, 1, 62, 8, 55.

Now we wish to ask several questions.

(a) For which m are the $\mathfrak{M}_m$ cyclic?

(b) Which $\mathfrak{M}_m$ are isomorphic? Generally when we pass from m to $m + 1$, we obtain a totally different pattern, e.g., $m = 54, 55, 56, 57$. But $\mathfrak{M}_3 \rightleftarrows \mathfrak{M}_4$, $\mathfrak{M}_{15} \rightleftarrows \mathfrak{M}_{16}$, and, more spectacularly, $\mathfrak{M}_{104} \rightleftarrows \mathfrak{M}_{105}$.

(c) For which m are the cycle graphs three-dimensional; as in $m = 63$, and, even more intricate, in $m = 91$?

(d) We note definite lobal patterns. Thus $\mathfrak{M}_{57}$ has nine lobes of the same type of which $\mathfrak{M}_{21}$ has three, and $\mathfrak{M}_8$, one. Again, $\mathfrak{M}_{56}$ has three lobes of the type of which $\mathfrak{M}_{24}$ has one; and $\mathfrak{M}_{55}$ possesses five $\mathfrak{M}_{15}$-type lobes. We ask, what is the structure of the various types of lobes, and how many such lobes may a group have?

(e) Can we characterize $\mathfrak{M}_m$ by a formula? Given m, we wish to determine the *structure* of $\mathfrak{M}_m$ by an (easily computable) formula. We recall, in this connection, that the *structure* of $\mathfrak{M}_{17}$ is clear even before we compute a primitive root.

(f) If $\mathfrak{M}_m$ is cyclic there is an a of order $\phi(m)$ modulo m. But if $\mathfrak{M}_m$ is not cyclic what is the largest order possible within the group?

(g) If $\mathfrak{M}_m$ is cyclic there are $\frac{1}{2}\phi(m)$ quadratic residues, but if $\mathfrak{M}_m$ is not, how many are there?

(h) Finally we note, from group theory, that every group of order 4 is either isomorphic to $\mathfrak{M}_8$ or to the cyclic $\mathfrak{M}_5$. There are only two *abstract* groups of order 4. Of order 8, there are five abstract groups, with cycle graphs as follows:

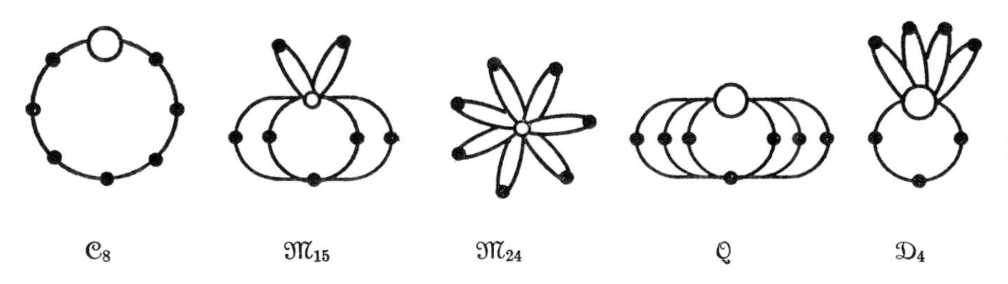

We see that two of them are isomorphic to $\mathfrak{M}_m$ groups. The cyclic group $\mathfrak{C}_8 \not\cong \mathfrak{M}_m$, but it is a *subgroup* of an $\mathfrak{M}_m$ group. Namely, $\mathfrak{C}_8$ is isomorphic to the group of quadratic residues of $\mathfrak{M}_{17}$—that is, to the cycle of 2 modulo 17 (see page 75).

The remaining two groups are well-known *non-Abelian* groups; $\mathfrak{Q}$ is the *quaternion* group, and $\mathfrak{D}_4$ is the *octic* group (the symmetries of a square). Since their multiplications are not commutative, they cannot be isomorphic to any $\mathfrak{M}_m$, or subgroup thereof. Therefore every Abelian group of order 8 is isomorphic to a subgroup of an $\mathfrak{M}_m$.

We now ask, is every finite Abelian group isomorphic to a subgroup of an $\mathfrak{M}_m$?

We close this section with a useful theorem.

Theorem 41. *In every finite Abelian group, if $x^2 = a$ possesses n solutions x, then every square, $y^2 = b$, possesses n solutions. In particular, in $\mathfrak{M}_m$, every quadratic residue has an equal number of square roots modulo m.*

PROOF. Let a have n square roots, $x_1, x_2, \cdots, x_n$. Let b have at least one, y_1. Then each element

$$y_i = y_1 x_1^{-1} x_i \tag{120}$$

for $i = 1, 2, \cdots, n$ satisfies $y_i^2 = b$ since $y_i^2 = y_1^2 x_1^{-2} x_i^2 = ba^{-1}a = b$.

Further, if $y_i = y_j$, we have $x_1 y_1^{-1} y_i = x_1 y_1^{-1} y_j$, and thus $x_i = x_j$. Therefore no square in the group can have fewer square roots than any other square.

It follows that if the cycle graphs for $\mathfrak{Q}$ and $\mathfrak{D}_4$ represent groups, (and they do), these groups cannot be Abelian, since in the octic group the identity has 6 square roots, while a second element has only 2. In the quaternion group the situation is reversed.

Exercise 63. Show that $\phi(m) = 8$ has exactly five solutions m, and that therefore $\mathfrak{M}_{24}$ is isomorphic to no other $\mathfrak{M}_m$.

Exercise 64. Each of the 7 rows in the table on page 47 form a subgroup of $\mathfrak{M}_{24}$ isomorphic to $\mathfrak{M}_8$.

Exercise 65. $\mathfrak{M}_{15}$ has both abstract groups of order 4 as subgroups.

Exercise 66. The quadratic residues of m constitute a subgroup of $\mathfrak{M}_m$. Call it $\mathfrak{Q}_m$. Then $\mathfrak{Q}_{55} \rightleftharpoons \mathfrak{M}_{11}$ and $\mathfrak{Q}_{65} \rightleftharpoons \mathfrak{M}_{21}$. But $\mathfrak{Q}_{63}$ is isomorphic to no $\mathfrak{M}_m$. Also $\mathfrak{Q}_{54} \rightleftharpoons \mathfrak{Q}_{57}$, etc.

Exercise 67. Draw a cycle graph for $\mathfrak{M}_{33}$.

Exercise 68. Determine the periods of the decimal expansions of $\frac{1}{57}$ and $\frac{1}{63}$ by examining the cycle graphs of $\mathfrak{M}_{57}$ and $\mathfrak{M}_{63}$.

Exercise 69. Determine 11^{-1}, 47^{-1}, and the four square roots of -1 modulo 65.

Exercise 70. Determine the order of 2 modulo 85. Interpret the result in terms of the equation $F_0 F_1 F_2 + 2 = F_3$. Compare Exercise 4.

Exercise 71. Let a finite group of order m contain a subgroup of order s. Then $s|m$. This is called *Lagrange's Theorem*—it generalizes Theorem 35.

Exercise 72. There is only one abstract group of a prime order—the cyclic group.

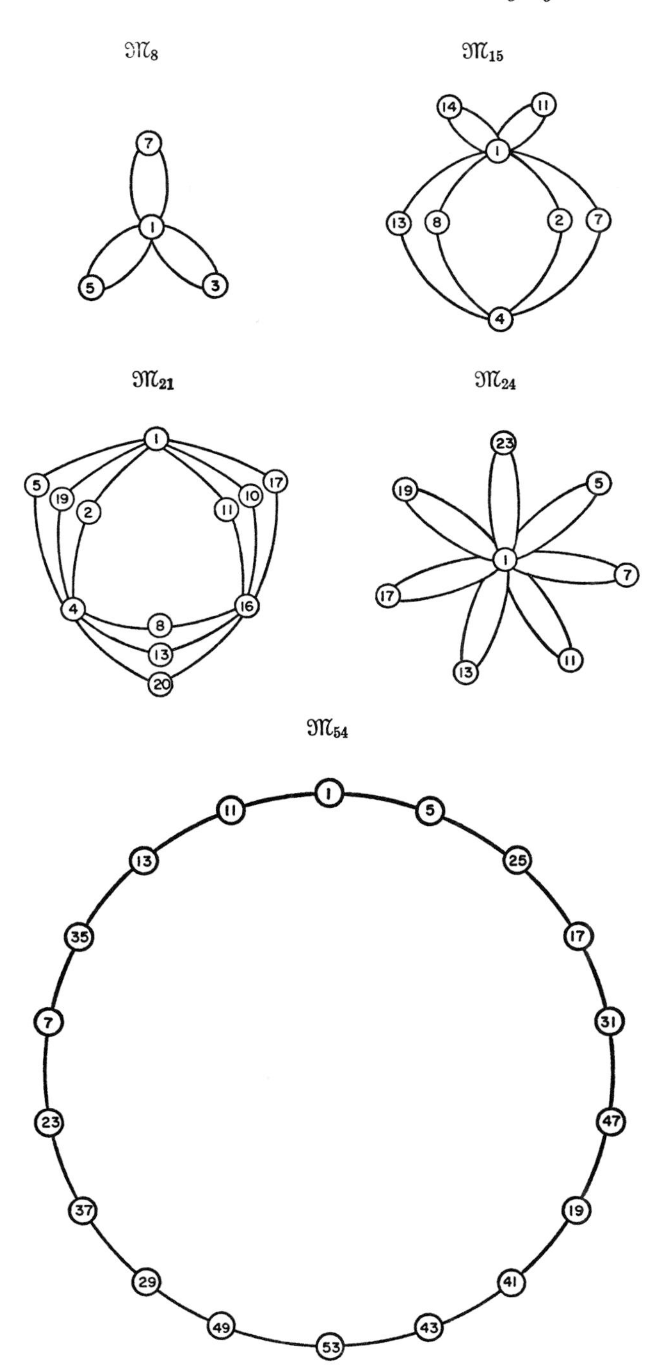

$\mathfrak{M}_8$
7
1
5
3
$\mathfrak{M}_{15}$
14
11
1
13
8
2
7
4
$\mathfrak{M}_{21}$
1
5
19
2
11
10
17
4
8
13
20
16
$\mathfrak{M}_{24}$
23
19
5
1
7
17
13
11
$\mathfrak{M}_{54}$
1
5
25
17
31
47
19
41
43
53
49
29
37
23
7
35
13
11

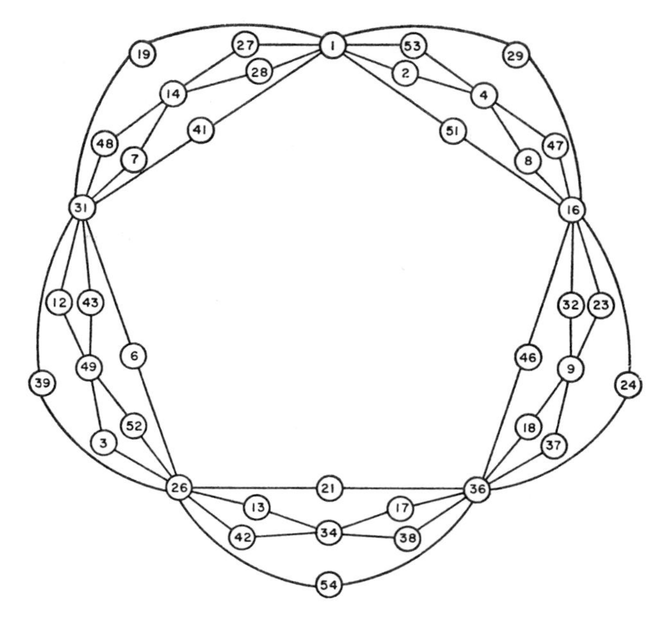

$\mathfrak{M}_{56}$

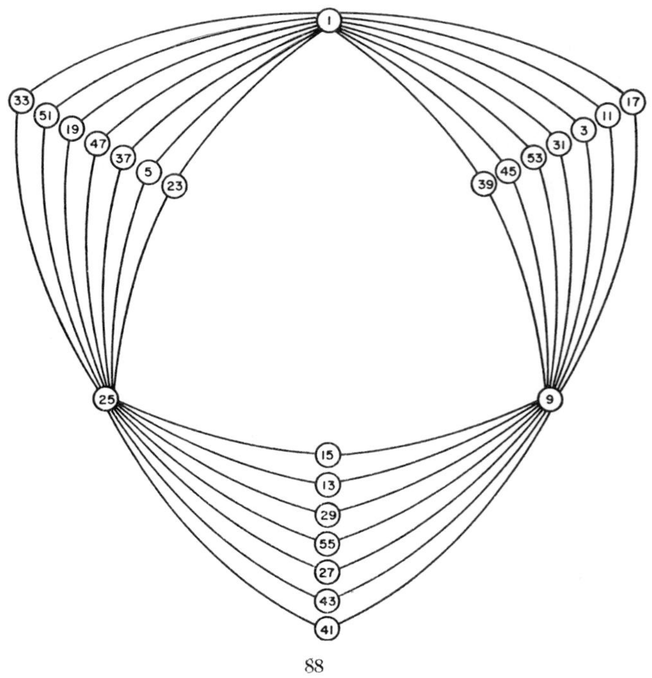

$\mathfrak{M}_{57}$

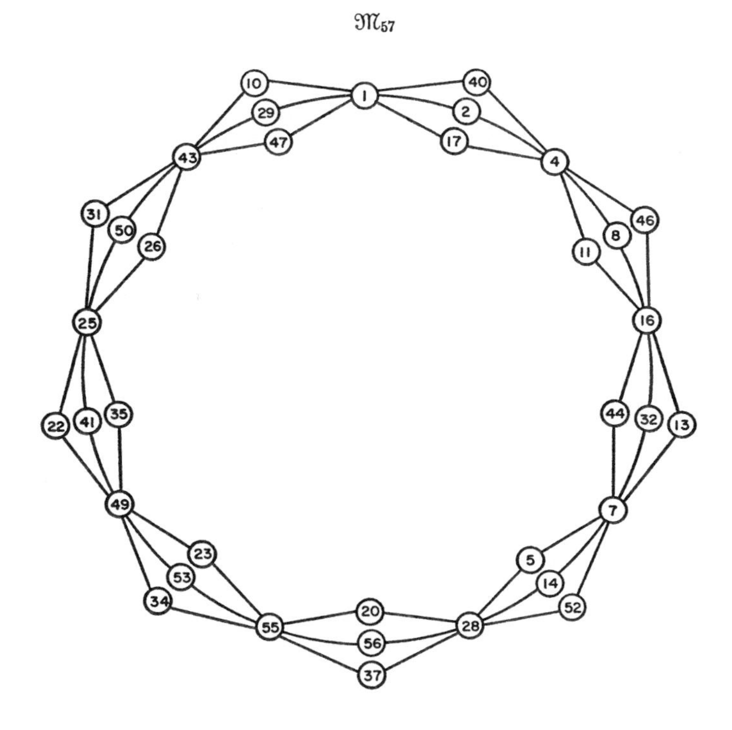

$\mathfrak{M}_{63}$

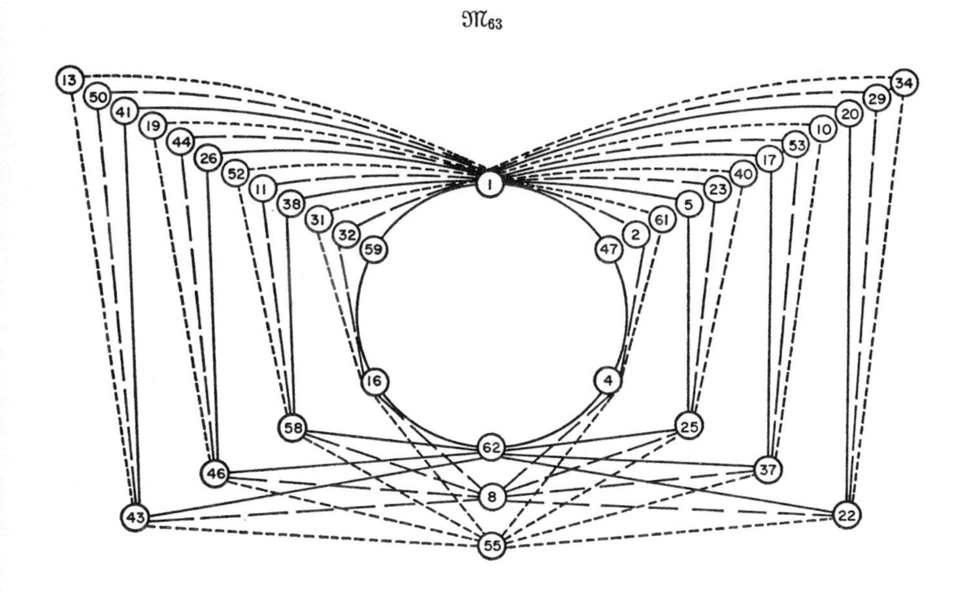

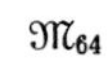

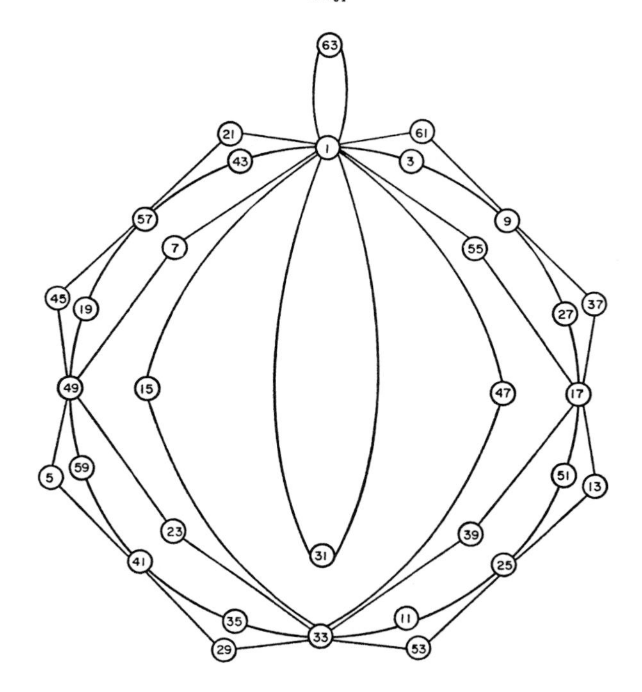

$\mathfrak{M}_{65}$

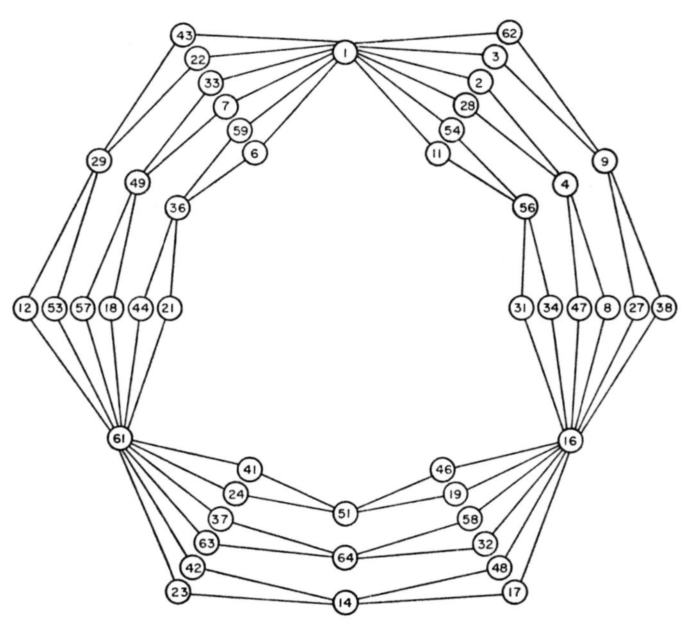

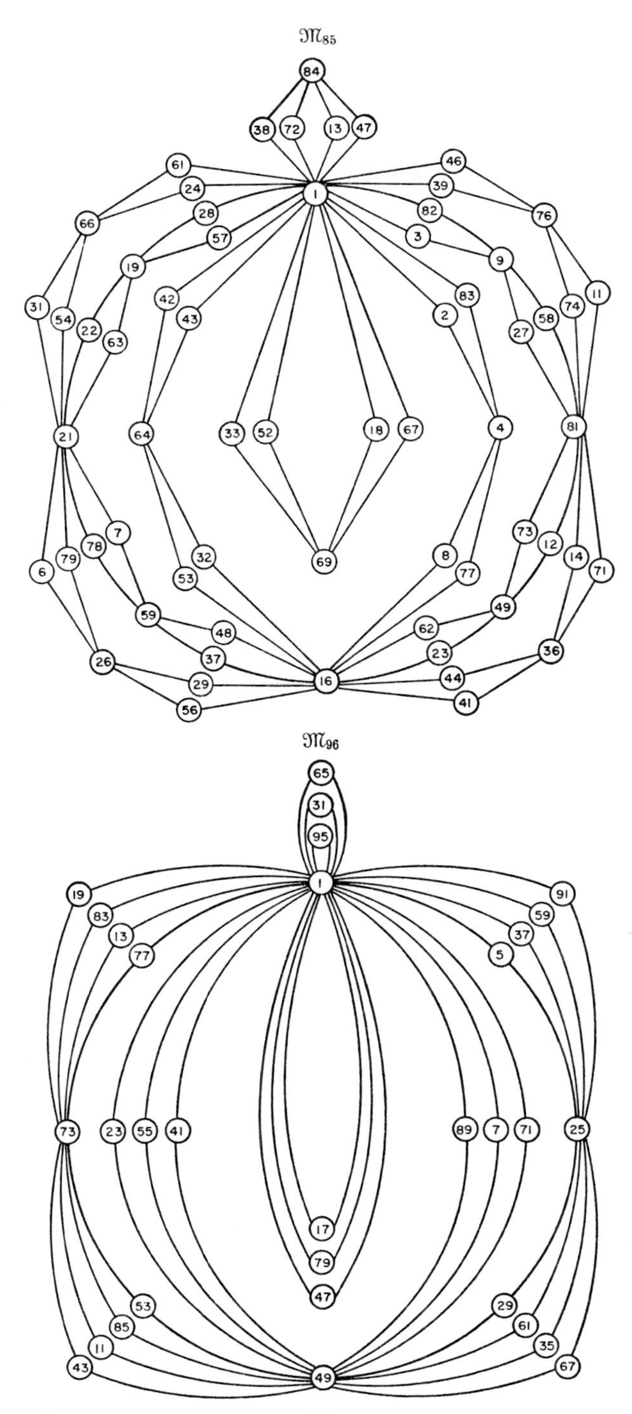

$\mathfrak{M}_{85}$
84
38 72 13 47
61 46
24 1 39
28 82
66 57 3 76
19 9
31 42 83 11
54 22 43 2 58 74
63 27
21 64 33 52 18 67 4 81
7 73 12
6 79 78 32 69 8 77 14 71
53
59 49
48 62
37 23 36
26 29 16 44
56 41
$\mathfrak{M}_{96}$
65
31
95
19 1 91
83 59
13 37
77 5
73 23 55 41 89 7 71 25
17
79
47
53 29
85 61
11 35
43 49 67

$\mathfrak{M}_{105}$

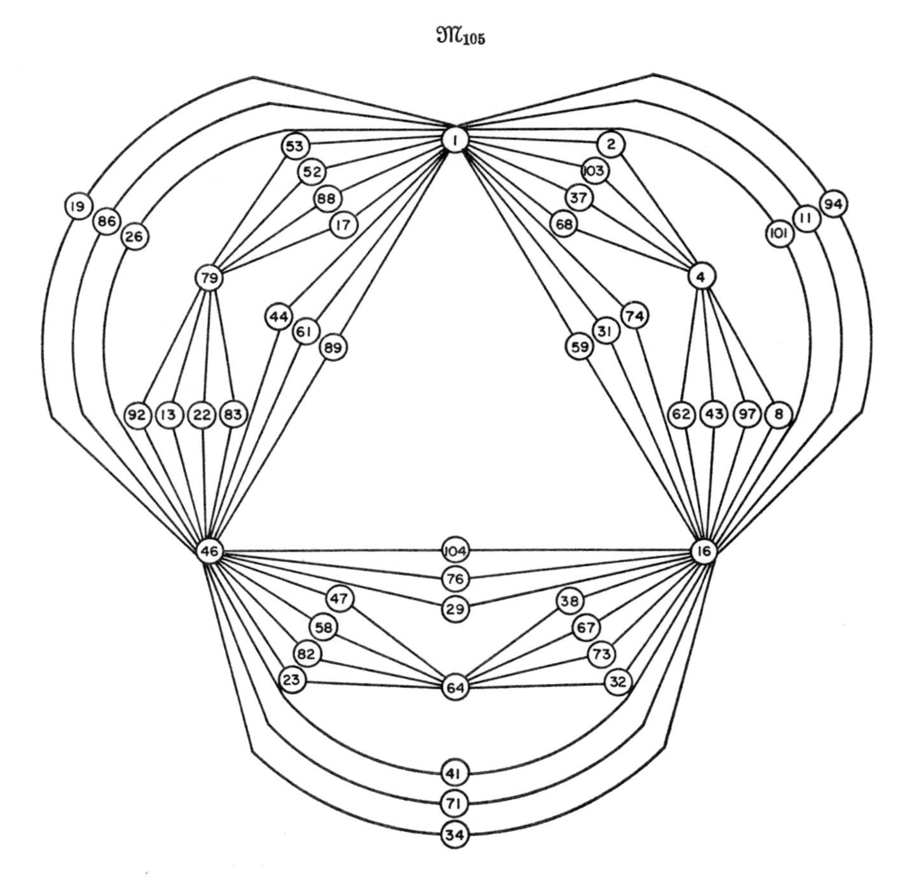

34. Answers Concerning Cycle Graphs

We shall prove

Theorem 42 (Gauss). *$\mathfrak{M}_m$ is cyclic—that is, m has a primitive root—if and only if m is one of the following:*

$$m = 2, 4, p^n, \quad or \quad 2p^n$$

where p is an odd prime and $n \geqq 1$.

Example.

$$\mathfrak{M}_{54} \text{ is cyclic, since } 54 = 2 \cdot 3^3.$$

Which $\mathfrak{M}_m$ are isomorphic? To answer this we need

Definition 29. By ϕ_m we mean a particular factorization of $\phi(m)$ obtained as follows if $m > 2$:

(A) Factor m into its standard form:

$$m = p_1^{a_1} p_2^{a_2} \cdots p_n^{a_n}. \tag{121}$$

(B) For each *odd* prime p_i write $\phi(p_i^{a_i}) = (p_i - 1)p_i^{a_i - 1}$ in a modified standard form

$$\phi(p_i^{a_i}) = \langle q_1^{b_1} \rangle \langle q_2^{b_2} \rangle \cdots \langle q_s^{b_s} \rangle \langle p_i^{a_i - 1} \rangle, \tag{122}$$

by factoring $p_i - 1$ into the prime powers $q_i^{b_i}$, and, if $a_i > 1$, by including the last factor. The symbol $\langle q_i^{b_i} \rangle$ means that the prime power is written as a single number, e.g., $\langle 5^2 \rangle = 25$.

(C) If $p_1 = 2$ and if $a_1 > 1$ we write $\phi(4) = 2, \phi(8) = 2 \cdot 2, \phi(16) = 2 \cdot 4$, and, in general,

$$\phi(2^{a_1}) = 2 \cdot \langle 2^{a_1 - 2} \rangle \qquad (a_1 \geqq 3). \tag{123}$$

If $a_1 = 1$, we omit this step.

(D) Now combine (C) and (B) into a modified standard factorization of $\phi(m)$:

$$\phi_m = 2 \cdot 2 \cdots 4 \cdot 4 \cdots 8 \cdots 3 \cdot 3 \cdots 9 \cdots 5 \cdots \tag{124}$$

Here $\phi(m)$ is factored into primes, and powers of primes, and we take care not to multiply factors of 2 with those of 4, etc.

If $m = 2$, we write $\phi_2 = 1$.

EXAMPLES:

$m = 105 = 3 \cdot 5 \cdot 7 \qquad \phi_{105} = 2 \cdot 2 \cdot 4 \cdot 3$

$m = 104 = 8 \cdot 13 \qquad \phi_{104} = 2 \cdot 2 \cdot 4 \cdot 3$

$m = 65 = 5 \cdot 13 \qquad \phi_{65} = 4 \cdot 4 \cdot 3$

$m = 15 = 3 \cdot 5 \qquad \phi_{15} = 2 \cdot 4$

$m = 16 \qquad \phi_{16} = 2 \cdot 4$

$m = 24 = 8 \cdot 3 \qquad \phi_{24} = 2 \cdot 2 \cdot 2$

$m = 63 = 9 \cdot 7 \qquad \phi_{63} = 2 \cdot 2 \cdot 3 \cdot 3$

$m = 17 \qquad \phi_{17} = 16$

Now we can state

Theorem 43. *$\mathfrak{M}_{m'}$ and $\mathfrak{M}_{m''}$ are isomorphic if and only if $\phi_{m'}$ and $\phi_{m''}$ are identical.*

EXAMPLES:

$$\mathfrak{M}_{104} \rightleftarrows \mathfrak{M}_{105} \not\rightleftarrows \mathfrak{M}_{65}$$

$$\mathfrak{M}_{15} \rightleftarrows \mathfrak{M}_{16} \not\rightleftarrows \mathfrak{M}_{24}$$

$$\mathfrak{M}_{40} \rightleftarrows \mathfrak{M}_{48} \not\rightleftarrows \mathfrak{M}_{17} \quad \text{(verify)}.$$

EXERCISE 73. If k is odd, $\mathfrak{M}_k \rightleftarrows \mathfrak{M}_{2k}$. If k is prime to 3 and 4, $\mathfrak{M}_{3k} \rightleftarrows \mathfrak{M}_{4k}$. If k is prime to 7 and 9, $\mathfrak{M}_{7k} \rightleftarrows \mathfrak{M}_{9k}$.

EXERCISE 74. Show that the $\mathfrak{M}_m$ are isomorphic for $m = 35, 39, 45, 52, 70, 78$, and 90.

EXERCISE 75. Show that the $\mathfrak{M}_m$ are isomorphic for $m = 51, 64, 68$, and 102, but for $m = 51, 80, 96$, and 120 we obtain 4 distinct abstract groups of the same order.

The last two theorems both follow from a more powerful result. To state this, it is convenient to modify the last definition to

Definition 30. By Φ_m we mean a particular factorization of $\phi(m)$ obtained as follows:

For each distinct prime q_i which divides $\phi(m)$ we take the largest power $\langle q_i^{b_i} \rangle$ which appears explicitly in ϕ_m and multiply these powers together. The product we call a *characteristic factor* of $\mathfrak{M}_m$. Setting this factor aside we repeat this operation with the remaining $\langle q_i^{b_i} \rangle$ in ϕ_m. Then Φ_m is the product of these characteristic factors

$$\Phi_m = f_1 \cdot f_2 \cdots f_r \tag{125}$$

where $f_1 \leqq f_2 \leqq f_3 \cdots \leqq f_r$.

EXAMPLES:

$$\Phi_{104,105} = 2 \cdot 2 \cdot 12$$

$$\Phi_{65} = 4 \cdot 12$$

$$\Phi_{15,16} = 2 \cdot 4$$

$$\Phi_{24} = 2 \cdot 2 \cdot 2$$

$$\Phi_{63} = 6 \cdot 6$$

$$\Phi_{17} = 16$$

Then we will have

Theorem 44. *If Φ_m is the product of r characteristic factors f_i, for each f_i there is a residue class g_i, of order f_i modulo m, such that every residue class a_j in $\mathfrak{M}_m$ can be expressed as*

$$a_j \equiv g_1^{s_{1,j}} g_2^{s_{2,j}} \cdots g_r^{s_{r,j}} \quad (mod\ m) \tag{126}$$

with

$$0 \leqq s_{i,j} < f_i$$

in one and only one way. We say that $\mathfrak{M}_m$ is the direct product *of the r cycles of the g_i.*

EXAMPLE:

For $m = 15$, $\Phi_{15} = 2 \cdot 4$, and we may take $g_1 = -1 \equiv 14 \pmod{15}$, of order 2, and $g_2 = 2$, of order 4. Then each of the 8 residue classes in $\mathfrak{M}_{15}$ is $\equiv (-1)^a 2^b \pmod{15}$ for one a and b such that $0 \leqq a < 2$, and $0 \leqq b < 4$.

The representation, Eq. (126), of $\mathfrak{M}_m$ as a direct product of r cycles is the characterization we sought in question (e) on page 85. We shall see presently that Theorems 42 and 43 are consequences of Theorem 44. But so are two others:

Theorem 45. *If f_i are characteristic factors of $\mathfrak{M}_m$, then*

$$f_i | f_j \tag{127}$$

if $i \leqq j$. It follows that if f_r is the largest characteristic factor of $\mathfrak{M}_m$,

$$a^{f_r} \equiv 1 \pmod{m} \tag{128}$$

for every residue class a in $\mathfrak{M}_m$.

COMMENT: Equation (128) gives us a sharpening of Euler's generalization of Fermat's Theorem. Further, it is clear from Eqs. (128) and (126) that f_r is the answer to question (f) on page 85.

Theorem 46. *If $m > 2$, and $\mathfrak{M}_m$ has r characteristic factors, m has $\phi(m) \cdot 2^{-r}$ quadratic residues, and each of these has 2^r square roots.*

EXAMPLE:

$$\mathfrak{M}_{105} \text{ has } \tfrac{48}{8} = 6 \text{ quadratic residues.}$$

PROOF OF THEOREM 45. Equation (127) is clear from the construction of the f_i in Definition 30. Then Eq. (128) follows at once from Eq. (126).

PROOF OF THEOREM 46. If $m > 2$, each contribution to ϕ_m, Eq. (122) in step B, and Eq. (123) in step C of Definition 29, is even. Therefore it follows that each f_i is even. It is then apparent, from Eq. (126), that a_j is a quadratic residue of m if and only if each of its exponents, $s_{i,j}$ is even. Since, by Theorem 41, each quadratic residue has an equal number of square roots, Theorem 46 follows.

PROOF OF THEOREM 42. If $m = 2, 4, p^n$, or $2p^n$ we find that $\Phi_m = \phi(m)$ with only one characteristic factor, and therefore g_1 is of order $\phi(m)$—that is, g_1 is a primitive root. Whereas if m is divisible by two distinct odd primes or equals $4k$ with $k > 1$, we find at least two characteristic factors. Since the largest, f_r, is less than $\phi(m)$, by Eq. (128) there is no primitive root.

PROOF OF THEOREM 43. First we note, by the construction, that $\phi_{m'}$ and $\phi_{m''}$ are identical if and only if $\Phi_{m'}$ and $\Phi_{m''}$ are identical. Then if $\phi_{m'}$ and $\phi_{m''}$ are identical, by the obvious mapping

$$g_1'^{s_1} g_2'^{s_2} \cdots g_r'^{s_r} \longleftrightarrow g_1''^{s_1} g_2''^{s_2} \cdots g_r''^{s_r}$$

we find that $\mathfrak{M}_{m'}$ and $\mathfrak{M}_{m''}$ are isomorphic. Conversely if they are isomorphic it is clear that $\phi(m') = \phi(m'')$ and also, from Theorem 46, $\mathfrak{M}_{m'}$ and $\mathfrak{M}_{m''}$ must have the same number of characteristic factors. We say further that $\Phi_{m'}$ and $\Phi_{m''}$ must in fact be identical, for, if not, we compare

$$\Phi_{m'} = f_1'\cdot f_2' \cdots f_r'$$

with

$$\Phi_{m''} = f_1''\cdot f_2'' \cdots f_r''$$

from right to left, and let $f_j' \neq f_j''$ be the largest factors which differ. Assume

$$F = f_j' < f_j'' = G$$

and let P be the product

$$f_{j+1}'\cdot f_{j+2}' \cdots f_r' = f_{j+1}''\cdot f_{j+2}'' \cdots f_r''.$$

Then the $R = \phi(m')/P$ residue classes,

$$g_1'^{s_1} g_2'^{s_2} \cdots g_j'^{s_j} \tag{129}$$

obtained by allowing the s_i to take on all values, all satisfy $x^F \equiv 1 (\text{mod } m')$. But all R of the residue classes

$$g_1''^{s_1} g_2''^{s_2} \cdots g_j''^{s_j}, \tag{130}$$

do not satisfy $x^F \equiv 1 \ (\text{mod } m'')$ since g_j'' is of order $G > F$. Let there be $S < R$ residues, Eq. (130), which do satisfy $x^F \equiv 1 \ (\text{mod } m'')$. All in all there are exactly RF^{r-j} solutions of $x^F \equiv 1 \ (\text{mod } m')$ since any of the R solutions of Eq. (129) may be multiplied by

$$g_{j+1}'^{s_{j+1}} g_{j+2}'^{s_{j+2}} \cdots g_r'^{s_r}$$

to yield another solution, if, and only if $f_{j+k} | s_{j+k}\cdot F$ for each k such that $j + 1 \leqq j + k \leqq r$. That is, each s_{j+k} can take on the F values

$$0, \quad \frac{f_{j+k}}{F}, \quad \frac{2f_{j+k}}{F}, \ldots, \frac{(F-1)f_{j+k}}{F}.$$

Likewise there are exactly SF^{r-j} solutions of $x^F \equiv 1 \ (\text{mod } m'')$. Since $S < R$ it follows that $\mathfrak{M}_{m'}$ and $\mathfrak{M}_{m''}$ are not isomorphic unless $\phi_{m'}$ and $\phi_{m''}$ are identical, since, in any isomorphism, 1 must map into 1, and the x' such that $x'^F \equiv 1$ must map into similar x''.

Theorem 44 is also one of the keys to the answer to the last question in Sect. 33, page 85. This answer is given by

Theorem 47. *Every finite Abelian group is isomorphic to a subgroup of* $\mathfrak{M}_m$ *for infinitely many different values of* m.

The two remaining questions in Sect. 33, (c) and (d), we shall here answer with less formality. We will state, without proof, that the cycle graph of $\mathfrak{M}_m$ is three dimensional if $\mathfrak{M}_m$ has at least two characteristic factors which are not powers of 2. Thus $\mathfrak{M}_m$ is three dimensional for $m =$ 63, 91, 275, and 341, since $\Phi_{63} = 6 \cdot 6$, $\Phi_{91} = 6 \cdot 12$, $\Phi_{275} = 10 \cdot 20$, and $\Phi_{341} =$ $10 \cdot 30$. See Exercise 19S, on page 206, for a sketch of the proof.

On the other hand, if

$$\Phi_m = \langle 2^a \rangle \cdot \langle 2^b \rangle \cdots \langle 2^y \rangle \cdot \langle 2^z N \rangle$$

where N is an odd number $\geqq 1$, the cycle graph will have N lobes, and each lobe is characterized by the formula $\{\langle 2^a \rangle \cdot \langle 2^b \rangle \cdots \langle 2^y \rangle \cdot \langle 2^z \rangle\}$. There are two different lobes of order 4:

the cyclic $\{4\}$ and $\{2 \cdot 2\}$.

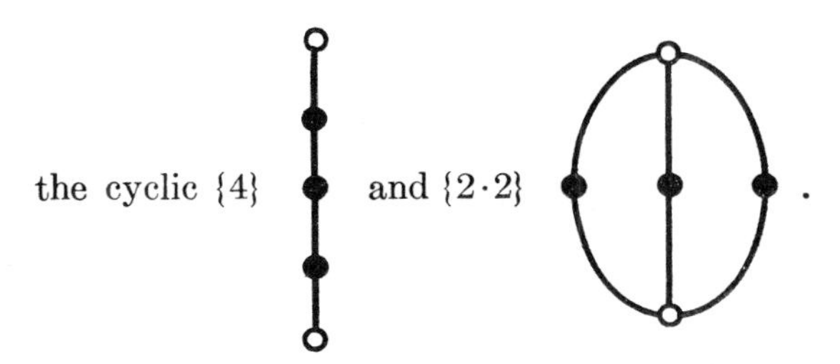

Thus the cyclic $\mathfrak{M}_{13}$ has 3 of the $\{4\}$, while $\mathfrak{M}_{21}$ (page 87) has 3 lobes $\{2 \cdot 2\}$.

There are three different lobes of order 8: the cyclic $\{8\}$; and

$\{2 \cdot 4\}$; and $\{2 \cdot 2 \cdot 2\}$

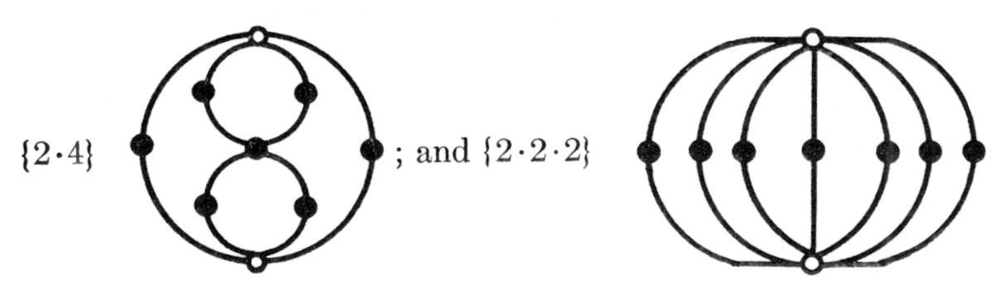

as in $\mathfrak{M}_{41}$, $\mathfrak{M}_{55}$ and $\mathfrak{M}_{56}$ respectively.

There are five different lobes of order 16: the cyclic $\{16\}$ in $\mathfrak{M}_{17}$; $\{2 \cdot 2 \cdot 4\}$ in $\mathfrak{M}_{105}$; $\{4 \cdot 4\}$ in $\mathfrak{M}_{65}$; $\{2 \cdot 8\}$ in $\mathfrak{M}_{32}$ (not shown); and $\{2 \cdot 2 \cdot 2 \cdot 2\}$ in $\mathfrak{M}_{168}$ (not shown).

How many different lobes are there of order 2^n? The answer is $p(n)$, the number of *partitions* of n. Thus $p(4) = 5$, since 4 may be partitioned (into positive integers) in five ways:

$$4 = 4$$

$$4 = 1 + 1 + 2$$

$$4 = 2 + 2$$

$$4 = 1 + 3$$

$$4 = 1 + 1 + 1 + 1$$

We will return to the theory of $p(n)$ in Volume II. To each partition of $n = n_1 + n_2 + \cdots + n_k$ there is a lobe of order 2^n:

$$\{\langle 2^{n_1}\rangle \cdot \langle 2^{n_2}\rangle \cdots \langle 2^{n_k}\rangle\}.$$

It will follow from Theorem 47 that for any such lobe, and for any *odd* N, there are infinitely many $\mathfrak{M}_m$ which have subgroups with a corresponding cycle graph.

But it is not possible to have two lobes of $\{2 \cdot 2\}$:

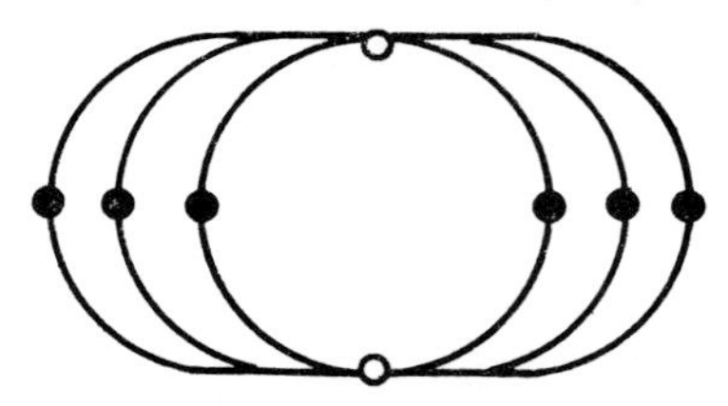

since we have seen that this group is non-Abelian (page 86). And four lobes of $\{2 \cdot 2\}$ does not represent *any* group—even a non-Abelian. It may be shown that it violates the associative law.

There remain the tasks of proving Theorems 44 and 47.

Exercise 76. Find the relationship between r, the number of characteristic factors of $\mathfrak{M}_m$, and the number of odd primes which divide m, and the power of 2 which divides m.

Exercise 77. For any a_j, in Eq. (126), which has every $s_{i,j}$ even, find explicitly its 2^r square roots.

Exercise 78. If $(a, 561) = 1$, then

$$a^{560} \equiv 1 \qquad (\text{mod } 561).$$

In particular, $561 | 2^{561} - 2$.

35. Factor Generators of $\mathfrak{M}_m$

We will prove Theorem 44 in three (rather long) steps.

Lemma 1. *Theorem 44 is true if m equals a prime power p^n. That is, if p is odd, or if $m = 2$ or 4, m has a primitive root. If $m = 2^n$ with $n \geqq 3$, $\Phi_m = 2 \cdot \langle 2^{n-2}\rangle$ and we have the representation*

$$a_j = (-1)^{s_j} 3^{t_j} \qquad (\text{mod } 2^n) \tag{131}$$

where -1 is of order 2, and 3 is of order 2^{n-2}.

Proof. We know that each p has a primitive root h. We first show that either h or $h + p$ is a primitive root of p^2. For, if $p > 2$, from Eq. (13) with $x = h + p$, $y = h$, and $n = p - 1$, we have

$$(h + p)^{p-1} - h^{p-1} = p[(h + p)^{p-2} + (h + p)^{p-3}h + \cdots + h^{p-2}]$$

and the square bracket has $p - 1$ terms, each of which is $\equiv h^{p-2} \pmod{p}$. But $p \nmid (p-1)h^{p-2}$ and we thus have

$$(h+p)^{p-1} \not\equiv h^{p-1} \pmod{p^2}.$$

Therefore at least one of the numbers h and $h + p$—call it g—satisfies

$$g^{p-1} \not\equiv 1 \pmod{p^2}. \tag{132}$$

By Theorem 35, if g is of order e modulo p^2, $e|\phi(p^2) = p(p-1)$. But, since g is of order $p - 1$ modulo p, we have $p - 1|e$. From Eq. (132), $e \neq p - 1$ and we therefore find $e = (p-1)p$, that is, g is a primitive root of p^2.

We thus have

$$g^{p-1} = 1 + kp \qquad (\text{with } p \nmid k).$$

By the binomial theorem

$$g^{(p-1)p} = 1 + kp^2 + (kp)^2 \frac{p(p-1)}{2} + tp^3$$

for some t, and, if $p > 2$,

$$g^{(p-1)p} = 1 + k_2 p^2 \qquad (\text{with } p \nmid k_2)$$

since

$$k_2 = k + \left(k^2 \frac{p-1}{2} + t\right) p.$$

By induction, for every *odd* p,

$$g^{(p-1)p^{s-1}} = 1 + k_s p^s \qquad (p \nmid k_s).$$

It follows, by the same argument as for p^2, that g is also a primitive root of p^s for every s. For

$$g^{(p-1)p^{s-2}} \not\equiv 1 \pmod{p^s}$$

and thus g is of order $(p-1)p^{s-1}$ modulo p^s.

For $p = 2$, we note

$$3^2 = 1 + 8$$

$$3^4 = 1 + 16 + 32t$$

$$3^8 = 1 + 32 + 64u$$

for integers t and u. By induction, 3 is of order 2^{n-2} modulo 2^n, if $n \geqq 3$. But none of the 2^{n-2} classes

$$a_j \equiv 3^{t_j} \pmod{2^n}$$

can be congruent to an

$$a_k \equiv -3^{t_k} \qquad (\text{mod } 2^n),$$

for, if so, we would have

$$8 | 3^a + 1$$

where $a = |t_j - t_k|$. This is not possible since $3^a + 1 \equiv 2$ or $4 \pmod 8$ for every a. Therefore the representation given by Eq. (131) gives every residue class in $\mathfrak{M}_{2^n}$.

On page 90 we see the cycles of 3 and $63 \equiv -1 \pmod{64}$. Each residue class -3^a has been placed close to $+3^a$.

Lemma 2. *If the* $\phi(A)$ *classes* a_i *in* $\mathfrak{M}_A$ *can be written*

$$a_i \equiv g_1^{\alpha_{1,i}} g_2^{\alpha_{2,i}} \cdots g_n^{\alpha_{n,i}} \qquad (\textit{mod } A) \tag{133}$$

where the factor generator g_j *is of order* m_j *modulo* A, *and*

$$m_1 m_2 \cdots m_n = \phi(A);$$

and if the $\phi(B)$ *classes* b_i *in a cyclic* $\mathfrak{M}_B$, *with* B *prime to* A, *are written*

$$b_i \equiv g^{\beta_i} \qquad (\textit{mod } B), \tag{134}$$

then the $\phi(AB)$ *classes* c_i *in* $\mathfrak{M}_{AB}$ *can be written*

$$c_i \equiv h_1^{\gamma_{1,i}} h_2^{\gamma_{2,i}} \cdots h_n^{\gamma_{n,i}} h^{\gamma_i} \qquad (\textit{mod } AB) \tag{135}$$

where the factor generator h_j *is of order* m_j, *and* h *is of order* $\phi(B)$, *modulo* AB.

Proof. Let

$$\begin{aligned} k &\equiv B^{-1}(1 - g) \qquad (\text{mod } A) \\ &\text{and} \quad k_j \equiv A^{-1}(1 - g_j) \qquad (\text{mod } B) \end{aligned} \tag{136}$$

for $j = 1, 2, \cdots, n$. Then set $h = Bk + g$, $h_j = Ak_j + g_j$, and we have

$$h = Bk + g \equiv 1 \qquad (\text{mod } A),$$

$$h_j = Ak_j + g_j \equiv 1 \qquad (\text{mod } B), \; (j = 1, 2, \cdots, n). \tag{137}$$

We now say that Eq. (135) has the stated properties. For, since $h_j \equiv g_j \pmod A$, h_j is of order m_j modulo A. Therefore

$$h_j^{m_j} = 1 + sA \quad \text{and also} \quad = 1 + tB.$$

Therefore $sA = tB$ and, since $(A, B) = 1$, we have $B | s$. Thus h_j is also of order m_j modulo AB, since, if it were of a smaller order modulo AB, this would imply a smaller order modulo A. Likewise h is of order $\phi(B)$ modulo AB.

For any *fixed* γ_i in Eq. (135), since $h^{\gamma_i} \equiv 1 \pmod{A}$, the $\phi(A)$ values of c_i are $\equiv$ to the $\phi(A)$ values of $a_i \pmod{A}$, and therefore are incongruent modulo A. On the other hand, for two different values of γ_i, the c_i are incongruent modulo B. Therefore each of the $\phi(A)\phi(B)$ values of c_i are incongruent modulo AB, since they are either incongruent modulo A, or modulo B, or both. Since each c_i is prime to both A and B, each is prime to AB, and since $\phi(AB) = \phi(A)\phi(B)$, the Lemma is proven.

Therefore, given any

$$m = p_1^{a_1} p_2^{a_2} \cdots p_n^{a_n},$$

we can construct a representation of $\mathfrak{M}_m$ in the form Eq. (135) by treating each $p_i^{a_i}$ by Lemma 1, and compounding them by Lemma 2. There remains the problem of putting the representation into the characteristic factor form, Eq. (126), of Theorem 44.

Lemma 3. *If g is of order AB modulo m with $(A, B) = 1$, the AB residue classes g^c $(0 \leqq c < AB)$ can be written as a direct product*

$$g^c \equiv s^a t^b \qquad (mod\ m) \tag{138}$$

where s is of order A and t is of order B modulo m. Conversely, given two residue classes s and t, of orders A and B, with $(A, B) = 1$, the AB classes on the right of Eq. (138) are all distinct, and constitute the cycle of some g of order AB modulo m.

Proof. Given g, set $s \equiv g^B$ and $t \equiv g^A$. Then s is of order A, t of order B, and the AB classes

$$s^a t^b \equiv g^{aB+bA} \qquad (0 \leqq a < A) \quad (0 \leqq b < B)$$

are distinct by Theorem 33. Conversely, given s and t, consider the right side of Eq. (138). Now if

$$s^{a_1} t^{b_1} \equiv s^{a_2} t^{b_2},$$

we have

$$s^{a_1 - a_2} \equiv t^{b_2 - b_1},$$

and

$$1 \equiv s^{(a_1 - a_2)A} \equiv t^{(b_2 - b_1)A}.$$

Then $B | (b_2 - b_1)A$, and, since $(B, A) = 1$, we have $b_1 = b_2$. Likewise $a_1 = a_2$. Thus the AB classes on the right of Eq. (138) are distinct. Now set $g \equiv st \pmod{m}$, and if g is of order e, then

$$s^e t^e \equiv 1.$$

Thus, as before, $A | e$, and $B | e$. Therefore $e = AB$. Further if $A^{-1}A \equiv 1$

(mod B), we have

$$A^{-1}A = kB + 1, \quad \text{and} \quad A^{-1}(b - a)A + a = k(b - a)B + b.$$

Let

$$f \equiv A^{-1}(b - a)A + a \qquad (\text{mod } AB)$$

or

$$f \equiv k(b - a)B + b \qquad (\text{mod } AB).$$

Therefore $g^f \equiv s^f t^f \equiv s^a t^b$ (mod m) so that the cycle of g contains the AB residue classes, and no others.

Thus, given any representation of $\mathfrak{M}_m$, obtained by Lemmas 1 and 2, we may decompose the cycles into cycles of prime-power order, as in ϕ_m, and then recompose them into the characteristic factors, as in Φ_m. This completes Theorem 44, and therefore also Theorems 42, 43, 45 and 46.

Previously we made the point that a primitive root for a prime modulus was proven to exist nonconstructively. We should now add that the *subsequent* steps in proving Theorem 44—that is, the foregoing three Lemmas—are all constructive, and involve explicitly given computations.

We note that a representation of $\mathfrak{M}_m$ in the form of Eq. (135) is not necessarily unique, even as to the number of factors, and can involve as many generators as the number of factors in ϕ_m, or as few generators as the number of factors in Φ_m.

We may also note that the last Lemma can assist us in the finding of primitive roots. Thus 2 is of order 3 modulo 7 and -1 is of order 2. Therefore $-2 \equiv 5$ is of order 6, that is, 5 is a primitive root, etc.

Exercise 79. A primitive root of p which is *not* a primitive root of p^2 is hard to come by. Show that 10 is a primitive root of 487 but not of 487^2 by computing $(10|487)$, and with reference to the congruences:

$$\begin{array}{llll}
100^3 \equiv 189 & (\text{mod } 487) & 100^3 \equiv 51324 & (\text{mod } 487^2) \\
189^3 \equiv 475 & & 51324^3 \equiv 100797 & \\
475^3 \equiv 220 & & 100797^3 \equiv 145833 & \\
220^3 \equiv 232 & & 145833^3 \equiv 78152 & \\
232^3 \equiv 1 & (\text{mod } 487) & 78152^3 \equiv 1 & (\text{mod } 487^2)
\end{array}$$

Find a primitive root of 487^2. Determine the periods of the decimal expansions of 487^{-1} and 487^{-2}.

Exercise 80. Find the fallacy in the following: If g is a primitive root both of p and of p^2, then every primitive root of p has the same property.

"Proof." Any primitive root h of p may be written $h = g^k$ where k is prime to $p - 1$, and $k < p - 1$. But

$$g^{p-1} = 1 + sp \qquad (p \nmid s).$$

Therefore $h^{p-1} = g^{(p-1)k} = 1 + ksp + tp^2$, or, since $p \nmid ks$, we may write

$$h^{p-1} = 1 + up \qquad (p \nmid u).$$

Therefore h is of order $p(p - 1)$ modulo p^2.

EXERCISE 81. Given g, a primitive root of p^n, with p odd, find a primitive root of $2p^n$.

EXERCISE 82. Determine a representation, Eq. (135), of $\mathfrak{M}_{35}$ by Lemmas 1 and 2. It will be a product of two cycles of orders 4 and 6. Now decompose and recompose into a product $2 \cdot 12$ and thus map $\mathfrak{M}_{35}$ isomorphically into $\mathfrak{M}_{39}$.

EXERCISE 83. Investigate the degeneration of Eq. (131) into *one* characteristic factor for $2^n = 2$ or 4. Note: $-1 \equiv 3 \pmod 4$.

EXERCISE 84. Show that $4 \nmid x^2 + 1$, and therefore $\sqrt{-1}$ does not exist modulo 2^n for $n > 1$. But if $\mathfrak{M}_{2^n}$ were cyclic, $\sqrt{-1}$ would exist if $n \geqq 3$. Thus 2^n has no primitive root if $n \geqq 3$.

EXERCISE 85. Let $n \geqq 3$. Show that r is a quadratic residue of 2^n if, and only if, $r = 8k + 1$. Thus 17 is the smallest positive integer, not equal to a square, which is a quadratic residue of 2^n. Note that in the cycle graph on page 90 the quadratic residues of 64 are strung out in numerical order! What do you make of that? Also, the smallest positive a for which $x^2 + a$ is divisible by *every* power of 2, for some x, is $a = 7$.

EXERCISE 86. For $m = 2^n$ with $n > 3$, show that the two classes of numbers $8k + 1$ and $8k - 1$ play special roles in the structure of $\mathfrak{M}_m$. But $8k + 3$ and $8k - 3$ play similar roles. How many subgroups of order 2^{n-2} are contained in $\mathfrak{M}_m$? Are they all isomorphic? Show that 3 may be replaced by 5 in Eq. (131).

EXERCISE 87. If p is an odd prime, r is a quadratic residue of p^n if and only if r is a quadratic residue of p.

EXERCISE 88. If $\mathfrak{M}_m$ is not cyclic, $a^{\frac{1}{2}\phi(m)} \equiv 1 \pmod m$ for every a prime to m. Thus Euler's Criterion can be generalized only to composites of the form $2p^n$ and p^n with p odd, and to 4. Further, if $\mathfrak{M}_m$ is represented by Eq. (126) the product A of *all* the residue classes a_i is given by

$$A \equiv (g_1 g_2 \cdots g_r)^{\frac{1}{2}\phi(m)} \pmod m,$$

and thus $A \equiv -1$ or $A \equiv +1$ according as $\mathfrak{M}_m$ is cyclic or not. Therefore (compare Exercise 25 on page 38) Wilson's Theorem, $p|(p-1)!+1$, like Euler's Criterion, only generalizes to these same composites (Gauss).

36. Primes in some Arithmetic Progressions and a General Divisibility Theorem

To prove Theorem 47 (page 96), we will assume, from group theory, that every finite Abelian group $\mathfrak{A}$ can be written as a *direct product* of cyclic subgroups. That is

$$a_i = g_1^{\alpha_{1,i}} g_2^{\alpha_{2,i}} \cdots g_n^{\alpha_{n,i}} \tag{139}$$

for every a_i in $\mathfrak{A}$. The generator g_j is of order m_j and the order of $\mathfrak{A}$ is the product $m_1 m_2 \cdots m_n$. This implies that the cycles of any two generators g_j and g_k have no element in common except the identity $g_j^0 = g_k^0$.

Now the reader may verify that Lemma 3 above holds for *every* finite Abelian group, so that any representation, Eq. (139), may be decomposed into cycles of prime-power order. Assume this done, and that m_j is now equal to $p_j^{\alpha_j}$ for p_j prime and $\alpha_j \geqq 1$.

Now let

$$N = q_1 q_2 \cdots q_n$$

where q_j is a prime of the form $kp_j^{\alpha_j} + 1$. Then $\mathfrak{M}_N$ will contain a cycle of order $q_j - 1 = kp_j^{\alpha_j}$ generated by a residue class s_j. Further $t_j \equiv s_j^k$ (mod N) has a cycle of order $p_j^{\alpha_j}$ and the *subgroup* of $\mathfrak{M}_N$ generated by

$$t_1^{\alpha_{1,i}} t_2^{\alpha_{2,i}} \cdots t_n^{\alpha_{n,i}}$$

is isomorphic to $\mathfrak{A}$.

Example:

Let $\mathfrak{A}$ be an Abelian group of order 9 represented by

$$a = x^\alpha y^\beta$$

where x and y are elements of $\mathfrak{A}$ both being of order 3. Then $\mathfrak{A}$ is isomorphic to a subgroup of $\mathfrak{M}_{91}$ since $91 = 7 \cdot 13$ and $7 = 2 \cdot 3 + 1$ while $13 = 4 \cdot 3 + 1$. Specifically, starting with 3 and 2 as primitive roots of 7 and 13 respectively, and using Lemma 2 with $A = 7$, $g_1 = 3$, $B = 13$, $g = 2$, etc., we obtain a representation of $\mathfrak{M}_{91}$ as

$$a \equiv 66^\alpha 15^\beta \pmod{91}$$

with 66 of order 6, and 15 of order 12. Then $66^2 \equiv 79$, and $15^4 \equiv 29$ are both of order 3, and $\mathfrak{A}$ is isomorphic to the subgroup of $\mathfrak{M}_{91}$ given by:

$$a \equiv 79^\alpha 29^\beta \pmod{91}.$$

(Note that $\mathfrak{A}$ is also isomorphic to the subgroup of quadratic residues $\mathfrak{Q}_{63}$, but this mapping is not obtained by the construction given above.)

But Eq. (139) may have an arbitrarily large number of factors of the same order, and therefore Theorem 47 follows if, and only if, there are infinitely many primes of the form $kp_j^{\alpha_j} + 1$ for *every* prime power $p_j^{\alpha_j}$. This is a special case of Dirichlet's Theorem 15 (page 22). But we have not proven Theorem 15.

Special cases of Dirichlet's Theorem may be proven by variations on the proof of Euclid's Theorem 8. Thus if M is a product of primes of the form $4k - 1$, $4M - 1$ must be divisible by a different prime of that form. For if every prime divisor of $4M - 1$ were of the form $4k + 1$, so would their product be of that form. With a similar definition of M, we may use $6M - 1$ to prove that there are infinitely many primes of the form $6k - 1$.

Using our knowledge of quadratic residues, we may similarly show that

$$M^2 + 2, \quad M^2 - 2, \quad \text{and} \quad M^2 + 4$$

must contain at least one new prime of the form $8k + 3$, $8k - 1$, and $8k - 3$ respectively. Again

$$\tfrac{1}{2}(M^2 + 1) \quad \text{and} \quad \tfrac{1}{4}(M^2 + 3)$$

are divisible *only* by primes of the form $4k + 1$ and $6k + 1$ respectively.

But it is clear that by such individual attacks we can never prove Dirichlet's Theorem, since this encompasses infinitely many cases. For our present purpose we do not need Dirichlet's Theorem in its full glory: "There are infinitely many primes of the form $ak + b$ for *every* $(a, b) = 1$." It suffices if $b = 1$ and a is any prime power, and this we may obtain by a very useful generalization of Fermat's Theorem 11.

Theorem 48. *Let $m = p^{n-1}$ with p prime and n positive. Let $(a, b) = 1$, $x = a^m$, $y = b^m$, and*

$$z = \frac{x^p - y^p}{x - y} = x^{p-1} + x^{p-2}y + \cdots + xy^{p-2} + y^{p-1}. \tag{140}$$

Then

$$(x - y, z) = 1 \quad \text{or} \quad p \tag{141}$$

according as

$$(a - b, p) = 1 \quad \text{or} \quad p.$$

Secondly,

$$p \mid z \quad \text{or} \quad p \nmid z \tag{142}$$

according as

$$p \mid (a - b) \quad \text{or} \quad p \nmid (a - b).$$

Thirdly, all other prime divisors of z are of the form $kp^n + 1$.

Before we prove Theorem 48 we shall give several applications.

(a) If $a = b + 1$ we find that *all* divisors of

$$z = \frac{(b+1)^{mp} - b^{mp}}{(b+1)^m - b^m} \tag{143}$$

are of the form $kp^n + 1$.

(b) If, in (a), $b = 1$, $n = 1$, and $m = 1$, we obtain Fermat's Theorem 11.

(c) If $a = 2s$, $b = 1$, $p = 2$, and $n - 1 = t$, we find that

$$z = \frac{(2s)^{2^{t+1}} - 1}{(2s)^{2^t} - 1} = (2s)^{2^t} + 1 \tag{144}$$

has divisors only of the form $k2^{t+1} + 1$.

(d) In particular, $t = 1$,

$$(2s)^2 + 1$$

is divisible only by divisors of the form $4k + 1$.

(e) And

$$(2s)^4 + 1$$

is divisible only by divisors of the form $8k + 1$.

(f) And the Fermat Number, obtained from Eq. (144) by $s = 1$ and $t = m$,

$$F_m = 2^{2^m} + 1,$$

has divisors only of the form $k2^{m+1} + 1$.

(g) If $a = 3s$, $b = 1$, $p = 3$, and $n = 1$, we find that

$$z = \frac{a^3 - 1}{a - 1} = a^2 + a + 1 = \tfrac{1}{4}[(2a + 1)^2 + 3]$$

$$= \tfrac{1}{4}[(6s + 1)^2 + 3]$$

has divisors only of the form $6k + 1$.

(h) If $p = 2$ and $n = 2$, then, if $(a, b) = 1$,

$$z = \frac{a^4 - b^4}{a^2 - b^2} = a^2 + b^2 \tag{145}$$

has only 2 and primes of the form $4k + 1$ as possible prime divisors.

(i) Finally we complete the

PROOF OF THEOREM 47. In Eq. (143) let $b = 1$. Then

$$\frac{2^{mp} - 1}{2^m - 1}$$

has at least one prime divisor of the form $p^n k + 1$. Given M, the product of a number of such primes, if $a = M$ and $b = M - 1$, we find from Eq. (140) that the z there contains at least one more. For every prime power p^n, there are therefore infinitely many primes of the form $p^n k + 1$. By the construction on page 104 there are therefore infinitely many $\mathfrak{M}_m$ with subgroups isomorphic to any finite Abelian group.

PROOF OF THEOREM 48. Let $g = (x - y, z)$. Then $y \equiv x \pmod{g}$, and, from Eq. (140), $z \equiv px^{p-1} \pmod{g}$. Thus $g | px^{p-1}$ and also $g | (x - y, px^{p-1})$. But since $(a, b) = 1$, we have $(x, y) = 1$, $(x - y, x) = 1$, and

$$(x - y, x^{p-1}) = 1.$$

Therefore $g = 1$ or p. Now, for every c,

$$c \equiv c^p \equiv c^{p^2} \equiv \cdots \equiv c^m \pmod{p}$$

by Fermat's Theorem. Therefore

$$x - y = a^m - b^m \equiv a - b \pmod{p},$$

and if $p \nmid (a - b)$, $p \nmid (x - y)$ and $g = 1$. But if $p | (a - b)$, $y \equiv x \pmod{p}$ and, by Eq. (140), $z \equiv px^{p-1} \pmod{p}$ or $p | z$ and $g = p$. This proves Eq. (141) and the first part of Eq. (142).

If q is a prime divisor of z, $q | x^p - y^p$ or $a^{p^n} \equiv b^{p^n} \pmod{q}$. Thus $q \nmid a$ and $q \nmid b$, for if it divided either, it would divide the other also, and this contradicts $(a, b) = 1$. Let b^{-1} satisfy $b^{-1}b \equiv 1 \pmod{q}$. Then

$$(b^{-1}a)^{p^n} \equiv 1 \pmod{q},$$

and since

$$(b^{-1}a)^{q-1} \equiv 1 \pmod{q}$$

by Fermat's Theorem, we obtain, by Theorem 10,

$$q | (b^{-1}a)^h - 1$$

where $h = (p^n, q - 1)$. If $h \neq p^n$, we must have $h | m = p^{n-1}$. Then

$$q | (b^{-1}a)^m - 1$$

or

$$q | (x - y).$$

But, by Eq. (141), q can then be only p, and that *only* if $p | (a - b)$. All other prime divisors of z, [that is, *all* prime divisors if $p \nmid (a - b)$], have $h = p^n$ and therefore are of the form $q = kp^n + 1$. This proves the third part of the theorem and the second part of Eq. (142).

With the foregoing theory we are now in position (in principle) to map any finite Abelian group isomorphically into a subgroup of an $\mathfrak{M}_m$ and therefore to carry out algebraic computations within the group by ordinary arithmetic. An example is given in the following exercise. We quote from a recent article in a digital computer newsletter.

EXERCISE 89. BINARY AND DECIMAL MACHINES AND ISOMORPHIC OPERATIONS.

"Certain operations, which are easy on binary machines, are awkward on decimal machines, and conversely. In particular, the logical AND, OR, and COMPLEMENT are naturals for binary machines while long numerical tables are often more quickly done on decimal machines since otherwise much machine time is used in binary-decimal conversion.

"Sometimes a very binaryish operation can nonetheless be done decimally by using isomorphic operations. To illustrate this, consider the following example.

"Let 'octal biconditional' be an operation which is designated by $*$ and which is performed on two (three-bit) binary numbers, from 000 to 111. Let

$$A*B = C$$

where A and B are two such numbers and the result C is a third. Then the first bit of C is a 1 if the first bits of A and B are equal. Otherwise, it is 0. The same rule holds for the second and third bits.

"Examples:

	$3*1 = 5$	(octal)
since	$011*001 = 101$	(binary).
Again	$5*4 = 6$	(octal)
since	$101*100 = 110$	(binary).

This operation, 'octal biconditional,' arose in a practical problem, namely, 'clipped autocorrelation.' It would seem to be very awkward to carry it out on a decimal machine.

"However, it is isomorphic to multiplication modulo 1000 according to the following mapping:

octal	0	1	2	3	4	5	6	7
	$\updownarrow$	$\updownarrow$	$\updownarrow$	$\updownarrow$	$\updownarrow$	$\updownarrow$	$\updownarrow$	$\updownarrow$
decimal	999	751	749	501	499	251	249	1

"For example, to compute

$$3*1$$

we may map 3 and 1 into 501 and 751 respectively, then multiply 501 and 751 *decimally*. The last three digits of the product are 251 and by mapping backward we find the answer, octal 5. Thus

$$3*1 = 5$$

as before."

Now the reader is asked to examine "octal biconditional" and, by comparing this with Φ_{1000}, to show that an isomorphic mapping such as that given follows from the theory above. Is there another mapping into $\mathfrak{M}_{1000}$ which does not use the same eight decimal numbers? Could we use $m = 100$ instead of $m = 1000$? What is the smallest modulus possible? Find a mapping for this modulus. From the remarks concerning lobal patterns on page 97 describe the cycle graph for $\mathfrak{M}_{1000}$. Where, in this pattern, are the eight decimal numbers utilized above?

Exercise 90. Find a prime of the form $9k + 1$ by the recipe given in the proof of Theorem 47. Find the two smallest primes of the form $3k + 1$ given by that recipe, and compare these with the two smallest primes of the same form which were used in the example on page 104.

Exercise 91. From a book on group theory or modern algebra obtain definitions of *quotient group* and *group of automorphisms*. Let $\mathcal{C}_\infty$ be the group of all integers under addition. Let $m > 0$, and let $\mathcal{C}_\infty^{(m)}$ be the multiples of m. Let $\mathfrak{A}_m$ be the group of m residue classes under *addition* modulo m. Then $\mathfrak{A}_m$ is the quotient group $\mathcal{C}_\infty/\mathcal{C}_\infty^{(m)}$. And $\mathfrak{M}_m$ is isomorphic to the group of automorphisms of $\mathfrak{A}_m$. And therefore every finite Abelian group is isomorphic to a subgroup of the group of automorphisms of a quotient group of an infinite cyclic group.

Can this characterization of Abelian groups—which seems to involve only group-theoretic concepts—be proven independently of the number-theoretic results in Theorem 48?

From the relationship between $\mathfrak{A}_m$ and $\mathfrak{M}_m$ explain the "coincidence" that the number of primitive roots of p and the order of $\mathfrak{M}_{p-1}$ both equal $\phi(p - 1)$.

37. Scalar and Vector Indices

If 3 is chosen as the primitive root of 17 we may have two tables:

i	0	1	2	3	4	5	6	7	8	9	10	11	12	13	14	15
a	1	3	9	10	13	5	15	11	16	14	8	7	4	12	2	6

a	1	2	3	4	5	6	7	8	9	10	11	12	13	14	15	16
i	0	14	1	12	5	15	11	10	2	3	7	13	4	9	6	8

In both tables

$$a \equiv 3^i \qquad (\text{mod } 17).$$

The exponent i is called the *index* of a modulo 17 and written

$$i = \text{ind } a. \tag{146}$$

Similar tables have been worked out for all moduli <2000 which are primes or powers of (odd) primes. They enable one to multiply, divide, and solve binomial congruences quite easily for these moduli. For example,

$$a_1 a_2 \equiv x \qquad (\text{mod } 17)$$

is solved by

$$\text{ind } x \equiv \text{ind } a_1 + \text{ind } a_2 \qquad (\text{mod } 16).$$

Thus, for

$$5 \cdot 6 \equiv x \qquad (\text{mod } 17)$$

$$\text{ind } x = 4 \equiv 5 + 15 \qquad (\text{mod } 16),$$

and therefore $x \equiv 13 \pmod{17}$. Similarly

$$ax \equiv b \qquad (\text{mod } 17)$$

is solved by

$$\text{ind } x \equiv \text{ind } b - \text{ind } a \qquad (\text{mod } 16).$$

With indices, as with logarithms, multiplication, division, evolution, and involution are replaced by addition, subtraction, multiplication, and division respectively. The general binomial congruence:

$$ax^n \equiv b \qquad (\text{mod } p), \tag{147}$$

is treated in

EXERCISE 92. If, in Eq. (147), n is prime to $p - 1$, there is a unique solution given by

$$\text{ind } x \equiv n^{-1}(\text{ind } b - \text{ind } a) \qquad (\text{mod } p - 1) \tag{148}$$

where n^{-1} is the reciprocal of n modulo $p - 1$.

If $(n, p - 1) = g$ and $g \nmid (\text{ind } b - \text{ind } a)$, there is no solution. But if $g | (\text{ind } b - \text{ind } a)$, there are g solutions given by

$$y + k\frac{p - 1}{g} \qquad (k = 0, 1, \cdots g - 1) \tag{149}$$

where

$$y \equiv \left(\frac{n}{g}\right)^{-1} \left(\frac{\text{ind } b - \text{ind } a}{g}\right) \quad \left(\text{mod } \frac{p-1}{g}\right). \tag{150}$$

EXERCISE 93. Solve

$$3x^6 \equiv 5 \qquad (\text{mod } 17).$$

If the modulus does not have a primitive root we must replace the *scalar* indices i with *vector* indices $(i, j, \cdots)$. For example, each of the 24 residue classes prime to 35 can, by the foregoing theory, be expressed as

$$a \equiv 8^i 26^j \qquad (\text{mod } 35)$$

with $i = 0, 1, \cdots, 3$ and $j = 0, 1, \cdots, 5$. When the vector index (i, j) has only 2 components a two-dimensional representation is handy. Thus

$i \backslash j$	0	1	2	3	4	5
0	1	26	11	6	16	31
1	8	33	18	13	23	3
2	29	19	4	34	9	24
3	22	12	32	27	2	17

(mod 35)

Then, as before, if

$$\text{ind } a = (i_1, j_1)$$

and

$$\text{ind } b = (i_2, j_2),$$

we have

$$\text{ind } ab = (i_3, j_3)$$

with $i_3 \equiv i_1 + i_2 \pmod{n_1}$ and $j_3 \equiv j_1 + j_2 \pmod{n_2}$ where the generators are of order n_1 and n_2 respectively. We may write:

$$\text{ind } ab \equiv \text{ind } a + \text{ind } b \quad (\text{mod } n_1, n_2).$$

That is, the indices are combined by *modulo vector addition*. Thus

$$33 \cdot 24 \equiv 22 \qquad (\text{mod } 35)$$

since

$$(3, 0) \equiv (1, 1) + (2, 5) \qquad (\text{mod } 4, 6).$$

Alternatively, we may consider the table to be continued periodically in both directions. Then ordinary vector addition suffices.

The problem of binomial congruences we leave to the reader.

We note that by the use of Lemma 3, page 101, the 4×6 table above can be transformed into a 2×12 table, etc. Even for a prime modulus, say 7, we may modify its one dimensional index into a two dimensional 2×3 diagram. Thus

1	2	4
6	5	3

(mod 7).

But a Fermat Prime, say 17, can *only* have a one dimensional representation since 2^{2^m} *cannot* be factored into two factors prime to each other. And an $\mathfrak{M}_m$ with 3 or more characteristic factors requires at least that many dimensions.

Finally we note that the (pattern of the) $\mathfrak{M}_m$ cycle graph is obtained most simply by the use of such modulo vector addition. Thus from

1	*A*	*B*	*C*
D	*E*	*F*	*G*

we at once obtain

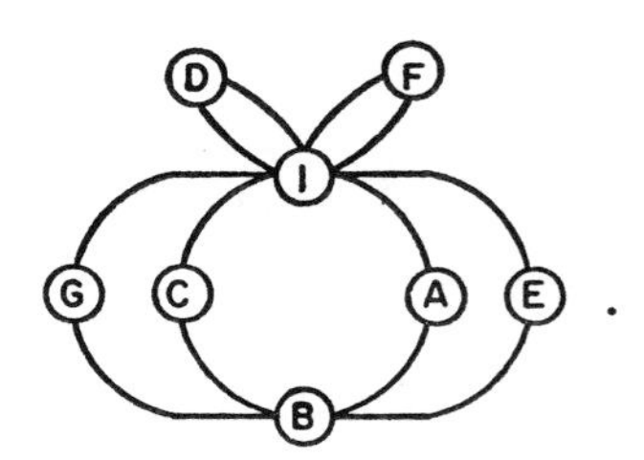

For the cycle of A is obtained by continued repetition of the vector displacement from 1 to A, giving us 1, A, B, C, and then, reducing the i coordinate modulo 4, back to 1. The continued repetition of the vector displacement from 1 to E, again reducing i by 4, or j by 2 when necessary, gives us 1, E, B, G, 1, etc. The elaborate pattern $\mathfrak{M}_{63}$ is most easily obtained not by *multiplication* modulo 63 but by *addition* of two dimensional vectors modulo (6, 6).

EXERCISE 94. Find the pattern of the cycle graphs for $\mathfrak{M}_{63}$ and for $\mathfrak{M}_{33}$ (Exercise 67) by the use of modulo vector addition.

EXERCISE 95. Show that if the octal numbers of Exercise 89 are subtracted from 7 and written in binary, they may be interpreted as vector indices of the corresponding decimal numbers.

EXERCISE 96. The transformation from a 4×6 representation to a 2×12 representation of $\mathfrak{M}_{35}$ (Exercise 82) may be interpreted as a linear transformation whereby a fundamental 4×6 rectangle becomes a fundamental 2×12 parallelogram.

38. THE OTHER RESIDUE CLASSES

After this detailed treatment of $\mathfrak{M}_m$ it is natural to ask "What of the residue classes *not* prime to m?" This can be answered quickly. Consider $m = 21$. Then besides the 12 solutions of $(x, m) = 1$, in $\mathfrak{M}_{21}$, there are 6 solutions of $(x, m) = 3$, 2 solutions of $(x, m) = 7$, and 1 solution of $(x, m) = 21$. These three sets of residue classes constitute three *other* groups under multiplication modulo 21. These groups have the cycle graphs

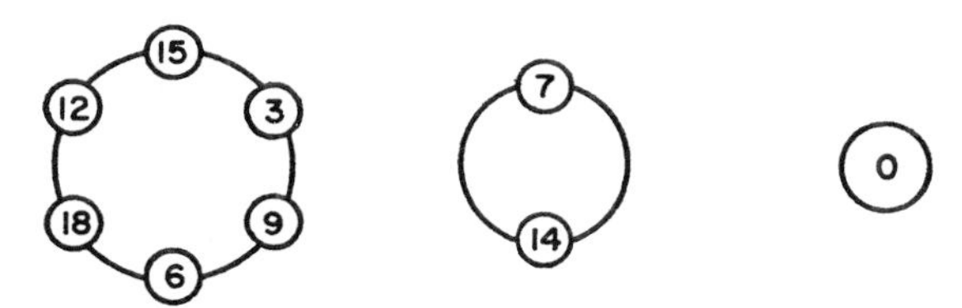

and the identities 15, 7, and 0 respectively. More generally we have

Theorem 49. *If $m = AB$ with $(A, B) = 1$, the $\phi(A)$ multiples of B, αB, where $(\alpha, A) = 1$, form a group under multiplication modulo m isomorphic to $\mathfrak{M}_A$. We call this group $\mathfrak{M}_m^{(B)}$. If β is the reciprocal of B modulo A, and*

$$\alpha\beta \equiv \bar{\alpha} \qquad (mod\ A), \tag{151}$$

the isomorphic mapping is

$$(mod\ A) \qquad \alpha \leftrightarrow \bar{\alpha}B \qquad (mod\ m), \tag{152}$$

and, in particular, βB is the identity of $\mathfrak{M}_m^{(B)}$.

EXAMPLE: Let $m = 21$, $A = 7$, and $B = 3$. Then $\beta \equiv 5 \pmod 7$ and 15 is the identity of $\mathfrak{M}_{21}^{(3)}$, as shown above. $\mathfrak{M}_{21}^{(3)} \rightleftarrows \mathfrak{M}_7$ under the mapping

$$(\text{mod } 7) \qquad \alpha \leftrightarrow 3\bar{\alpha} \qquad (\text{mod } 21)$$

where $\bar{\alpha} \equiv 5\alpha \pmod 7$.

PROOF. By Theorem 17 the $\bar{\alpha}$'s are a rearrangement of the α's. If

$$\alpha_1\alpha_2 \equiv \alpha_3 \qquad (\text{mod } A),$$

$$\bar{\alpha}_1 B\bar{\alpha}_2 B \equiv \alpha_1\alpha_2(\beta B)^2 \equiv \alpha_3\beta B \equiv \bar{\alpha}_3 B \qquad (\text{mod } A).$$

And clearly

$$\bar{\alpha}_1 B\bar{\alpha}_2 B \equiv \bar{\alpha}_3 B \qquad (\text{mod } B).$$

Therefore

$$m|\bar{\alpha}_1 B\bar{\alpha}_2 B - \bar{\alpha}_3 B \qquad \text{or}$$

$$\bar{\alpha}_1 B\bar{\alpha}_2 B \equiv \bar{\alpha}_3 B \pmod{m}.$$

Thus the $\phi(A)$ multiples of B prime to A form a group isomorphic to $\mathfrak{M}_A$ under multiplication modulo m.

Definition 31. An integer not divisible by a square greater than 1 is called *quadratfrei.*

Theorem 50. *If m is divisible by n distinct primes, there will be exactly 2^n multiplicative groups modulo m of the type $\mathfrak{M}_m^{(B)}$ described in Theorem 49. If m is quadratfrei, each of the m residue classes is contained in exactly one of these groups. If m is not quadratfrei, those residue classes a satisfying*

$$(a, m) = g \quad \textit{with} \quad (g, m/g) \neq 1 \tag{153}$$

are contained in no multiplicative group.

Proof. If

$$m = p_1^{a_1} p_2^{a_2} \cdots p_n^{a_n},$$

we may clearly choose the B, (and therefore the A), of Theorem 49 in 2^n different ways. If m is quadratfrei, each a_i equals 1, and therefore B may be any divisor of m. Since the residue classes in $\mathfrak{M}_m^{(B)}$ satisfy $(x, m) = B$, no residue class belongs to two of these groups, and, if m is quadratfrei, every possible greatest common divisor, $g = (x, m)$, occurs as a B. In this case, then, each residue class is in exactly one group.

But if one or more $a_i \neq 1$, and if $g = sp_i^{\alpha}$ with $1 \leqq \alpha < a_i$, and $p_i \nmid s$, let a be a residue class satisfying Eq. (153). Then $a = tp_i^{\alpha}$ with $p_i \nmid t$. It follows that

$$a^r \equiv a \pmod{p_i^{a_i}}$$

for no $r > 1$, and therefore

$$a^r \equiv a \pmod{m}$$

for no such r. But if a were in a group of order h, and that group had an identity e, we would have

$$a^h \equiv e \text{ and } a^{h+1} \equiv a \pmod{m}.$$

Thus if m is not quadratfrei there are still only 2^n groups, and all remaining residue classes, Eq. (153), are in no group.

Corollary. *There are exactly 2^n solutions of*

$$x^2 \equiv x \pmod{m}$$

if m is divisible by exactly n distinct primes.

Proof. Any such x is the identity of a multiplicative group modulo m.

Example: If $m = 36$ there are four $\mathfrak{M}_m^{(B)}$ isomorphic to the four $\mathfrak{M}_m^{(B)}$ for $m = 21$. ($\mathfrak{M}_m^{(1)}$ is now our former $\mathfrak{M}_m$.) The remaining 15 residue classes modulo 36 have *powers* in one of the $\mathfrak{M}_{36}^{(B)}$, although they themselves remain outside. We may diagram these appendages as follows:

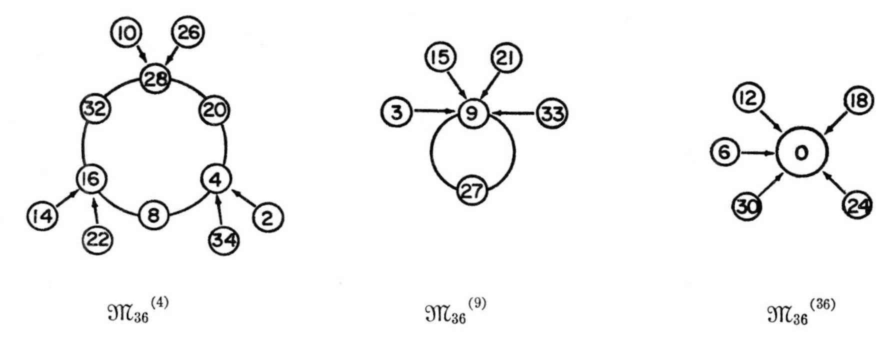

$\mathfrak{M}_{36}^{(4)}$ $\mathfrak{M}_{36}^{(9)}$ $\mathfrak{M}_{36}^{(36)}$

$\mathfrak{M}_m^{(1)}$ never has appendages. These extra residue classes "join" the group *irreversibly*. Their powers get in, but can't get out.

Let us also note

Theorem 51. *The 2^n multiplicative groups $\mathfrak{M}_m^{(B)}$ are isomorphic to subgroups of $\mathfrak{M}_m^{(1)} = \mathfrak{M}_m$.*

The proof is left to the reader.

Exercise 97. Interpret the proof of Theorem 1, on page 4, in terms of $\mathfrak{M}_{10}^{(2)}$.

39. The Converse of Fermat's Theorem

If N is a prime $\neq 2$,

$$2^{N-1} \equiv 1 \pmod{N}. \tag{154}$$

The converse is not true. Thus

$$2^{560} \equiv 1 \pmod{561}$$

as in Exercise 78, but $561 = 3 \cdot 11 \cdot 17$ is not a prime. The smallest composite N which satisfies Eq. (154) is $341 = 11 \cdot 31$. In fact

$$x \equiv 2^a 3^b \pmod{341} \tag{155}$$

is a representation of $\mathfrak{M}_{341}$ where 2 is of order 10, and 3 is of order 30. So $2^{10} \equiv 1 \equiv 2^{340} \pmod{341}$.

Definition 32. A *fermatian* is an integer N which satisfies Eq. (154).

Definition 33. A *Fermat number* F_m is one of the form $2^{2^m} + 1$.

Definition 34. A *Carmichael number* m is a composite whose largest characteristic factor, f_r, divides $m - 1$. See Definition 30.

Definition 35. A *Wieferich Square* is the square of a prime p such that $p^2 | 2^{p-1} - 1$.

Wieferich Squares enter into the theory of Fermat's Last Theorem.

Theorem 52. *All odd primes, Fermat numbers, Mersenne numbers, Carmichael numbers and Wieferich Squares are fermatians. There are other fermatians, also, since* 341, *for instance, is none of these.*

PROOF. Odd primes are obviously fermatians. Since $F_m | 2^{2^{m+1}} - 1$, and $2^{m+1} | 2^{2^m}$, we find

$$F_m | 2^{F_m - 1} - 1.$$

Again

$$M_p = 2^p - 1 = kp + 1.$$

Then

$$M_p = 2^p - 1 | 2^{kp} - 1 = 2^{M_p - 1} - 1.$$

A Carmichael number m must be odd, since f_r is even, (proof of Theorem 46), and thus could not divide $m - 1$ if m were even. Therefore, by Eq. (128),

$$2^{m-1} \equiv 1 \pmod{m}.$$

Also, if $p^2 | 2^{p-1} - 1$, $p^2 | 2^{p^2 - 1} - 1$. Finally 341 is none of these. It is not a Carmichael number since $f_r = 30 \nmid 340$.

It has never been proved that

(a) There are infinitely many Mersenne composites, or
(b) There are infinitely many Fermat composites, or
(c) There are infinitely many Carmichael numbers, or
(d) There are infinitely many Wieferich Squares.

Of the last there are only two examples up to $p = 100{,}000$; (S. Kravitz). These are 1093^2 (Meissner) and 3511^2 (Beeger).

Nonetheless it is easy to prove

Theorem 53. *There are infinitely many composite fermatians.*

PROOF. Suppose f_1 is a composite fermatian. Then

$$f_2 = 2^{f_1} - 1$$

is also one. For if $f_1 | 2^{f_1 - 1} - 1$, $f_1 | f_2 - 1$, and $f_2 = 1 + kf_1$. Then

$$2^{f_2 - 1} - 1 = 2^{kf_1} - 1$$

which is divisible by $2^{f_1} - 1 = f_2$. Further, by Theorem 4, f_2 is composite if f_1 is. Since $2^{11} - 1$, say, is a composite fermatian, there are infinitely many of them.

The first ten composite fermatians are

$341 = 11 \cdot 31$	561 (a Carmichael)
$645 = 3 \cdot 5 \cdot 43$	1105 (a Carmichael)
$1387 = 19 \cdot 73$	1729 (a Carmichael)
$1905 = 3 \cdot 5 \cdot 127$	2047 (a Mersenne)
2465 (a Carmichael)	$2701 = 37 \cdot 73$

The 43rd composite fermatian, $31417 = 89 \cdot 353$, belongs to none of the foregoing distinguished classes, but is perhaps distinguished in its own right:

$$31417 \mid 2^{31416} - 1.$$

P. Poulet and D. H. Lehmer have tabulated all composite fermatians $<10^8$. We give a table showing $C(N)$, the number of such composites $\leqq N$. This is compared with $\pi(N)$.

N	$C(N)$	$\pi(N)$	$\pi(N)/C(N)$
10	0	4	—
10^2	0	25	—
10^3	3	168	56
10^4	22	1229	56
10^5	79	9592	121
10^6	247	78498	318
10^7	750	664579	886
10^8	2043	5761455	2820

Apparently composite fermatians are relatively rare. Of these 2043 composites we may note that 252 are Carmichaels, 2 are Mersennes, 2 are Wieferich Squares, and none are Fermat numbers. For the entries in the table we have

$$C(N) < \sqrt{\pi(N)}. \tag{156}$$

Definition 36. If a class of positive integers A contains a subclass B, and if A and B are equinumerous, we say *almost all* A numbers are B numbers.

Example: From the prime number theorem, almost all positive integers are composite.

While it has not been proven that Eq. (156) remains valid as $N \to \infty$, one is tempted to risk

Conjecture 15. *Almost all fermatians are prime.*

Composite fermatians have some interesting properties (Poulet). Their distribution is very irregular. Thus 65,350,801 and 65,350,821 are successive composite fermatians, and so are 95,452,781 and 96,135,601—a gap of 20, and another of more than a half of a million. Very unexpected is the fact that more than one half of these numbers end in the digit 1.

EXERCISE 98. Prove that 1105 is a Carmichael number.

EXERCISE 99. The divisibility relation defining Wieferich Squares reminds one of the rare primitive roots of p which are not primitive roots of p^2. But show that 2 is not a primitive root of 3511. Nor is it of 1093, but that is not as easy.

40. Sufficient Conditions for Primality

When we left the perfect numbers we were in need of a good criterion for the primality of M_p. Wilson's Theorem:

$$(N - 1)! \equiv -1 \pmod{N}, \tag{157}$$

is a necessary and sufficient condition, but it is not practical. Fermat's Theorem:

$$2^{N-1} \equiv 1 \pmod{N}, \tag{158}$$

is a necessary and practical condition, but it is not sufficient—as we have just seen. We may even say that it is particularly useless for Mersenne and Fermat numbers, in view of Theorem 52. This is unfortunate, for while 2^{N-1}, like $(N - 1)!$, also grows rapidly, it is relatively easy to compute—by successive squarings and residue arithmetic.

We note that while $N = 341$ passes the test of Eq. (158), it does not pass the test:

$$3^{340} \equiv 1 \pmod{341},$$

since, by Eq. (155), 3 is of order 30 and thus $3^{340} \equiv 3^{10} \not\equiv 1 \pmod{341}$.

But a Carmichael number m passes the test

$$a^{m-1} \equiv 1 \pmod{m}$$

for every a prime to m. Because of this these numbers are also called *pseudoprimes*. By the results of Sect. 38 we may state an even stronger result.

Theorem 54. *For every Carmichael number m, and any a,*

$$m | a^m - a; \tag{159}$$

just as in Fermat's Theorem 13:

$$p | a^p - a.$$

COMMENT: By implication the test is truly infantile since the number doesn't know its $m|a^m - a$ from its $p|a^p - a$.

PROOF. A Carmichael number m is quadratfrei, for if $p^2|m$, we have $p|\phi(m)$, and therefore $p|f_r$, its largest characteristic factor. But if $m = ps$, and $f_r = pt$, we see that $f_r \nmid m - 1$. Now, since m is quadratfrei, by Theorems 50 and 51, every residue class a modulo m is in an $\mathfrak{M}_m^{(B)}$ isomorphic to a subgroup of $\mathfrak{M}_m$. Thus

$$a^{f_r} \equiv I \qquad (\text{mod } m)$$

where I is the identity of $\mathfrak{M}_m^{(B)}$. Then $a^{m-1} \equiv I$ and $a^m \equiv a \pmod{m}$.

We now seek a better criterion and we decide that Euler's Criterion is twice as good a test as Fermat's Theorem. If 341 were a prime, since it is of the form $8k + 5$, we would have $(2|341) = -1$, and

$$2^{170} \equiv -1 \qquad (\text{mod } 341).$$

But since $2^{10} \equiv 1$ and $2^{170} \equiv 1$, we see that 341 does not pass this test. If a composite N passes Eq. (158), it may be expected to pass

$$2^{(N-1)/2} \equiv (2|N) \qquad (\text{mod } N) \tag{160}$$

only one-half the time. Here the "Legendre symbol," $(2|N)$, is computed as if N were a prime. Nonetheless, Eq. (160) is not sufficient either, and, in particular, all Mersenne numbers satisfy this congruence.

In contrast, Euler's Criterion, with a base 3, is a necessary *and* sufficient condition for the primality of Fermat numbers.

Theorem 55 (Pepin's Test). $F_m = 2^{2^m} + 1$ *is a prime if and only if*

$$3^{(F_m-1)/2} \equiv -1 \qquad (\textit{mod } F_m). \tag{161}$$

PROOF. In Theorem 40 we showed that if F_m is a prime, $(3|F_m) = -1$, and, by Euler's Criterion, Eq. (161) follows. The converse interests us more. If Eq. (161) is true, so is

$$3^{F_m-1} \equiv 1 \qquad (\text{mod } F_m).$$

Then if $p|F_m$, $3^{F_m-1} \equiv 1 \pmod{p}$, and the order of 3 modulo p divides $F_m - 1 = 2^{2^m}$. This order is thus a power of 2. But it cannot divide $2^{2^m-1} = (F_m - 1)/2$ since that would contradict Eq. (161). Therefore the order is $F_m - 1$, and since it must be $\leqq p - 1$, we have $F_m \leqq p$. Thus $p = F_m$ and Eq. (161) is also a sufficient condition for the primality of F_m.

The reader will hear a familiar ring in the argument. We use the fact that a divisor d of p^n, with p a prime, divides p^{n-1}, if it does not equal p^n. If this leads to a contradiction, d must equal p^n. In Theorem 55 $p = 2$, but in Theorem 48 p is any prime.

With this success for Fermat numbers we again inquire about Mersenne numbers, $2^p - 1 = M_p$. Here M_p again involves a power of 2, but this time $M_p - 1$ is not that power of 2. Instead $M_p + 1$ is. Here we see the difficulty. What we need are not divisibility theorems like Fermat's Theorem and Euler's Criterion, since these involve $N - 1$. We need a divisibility theorem involving $N + 1$. Lucas found such a theorem, and by the use of it he obtained the Lucas Criterion for Mersenne numbers.

The theorem is associated with rational *approximations* to the $\sqrt{3}$. When the $\sqrt{3}$, and earlier, the $\sqrt{2}$, were found to be irrational, there was a great crisis in Greek mathematics and philosophy. We close the present chapter, and start a new one, which discusses this crisis, and, associated with it, another important source of number theory.

Exercise 100. If $2^m + 1$ is prime, m is a power of 2.

Exercise 101. From case (f) of Theorem 48, page 106, if a prime $p|F_m$, $p = 1 + k2^{m+1}$. Show that 2 is of order 2^{m+1} modulo p, and also, that if $m > 1$, $(2|p) = 1$. Then $2^{(p-1)/2} \equiv 1 \pmod{p}$, and k is even. Thus $p = 1 + s2^{m+2}$ if $m > 1$.

Exercise 102. From Exercises 100, 101, and 4, if we search for the smallest prime which divides F_5, our first trial divisor is 641.

Exercise 103. Prove that every Mersenne number passes the Euler Criterion test, Eq. (160), as stated on page 119.

CHAPTER III

PYTHAGOREANISM AND ITS MANY CONSEQUENCES

41. The Pythagoreans

We now examine a third source of number theory, one much older than periodic decimals, and even older than perfect numbers.

Definition 37. *Pythagorean numbers* are three positive integers that satisfy the equation

$$a^2 + b^2 = c^2. \tag{162}$$

The name has a twofold significance. First, it refers to the Pythagorean Theorem concerning a right triangle, and the three integers give us such a triangle:

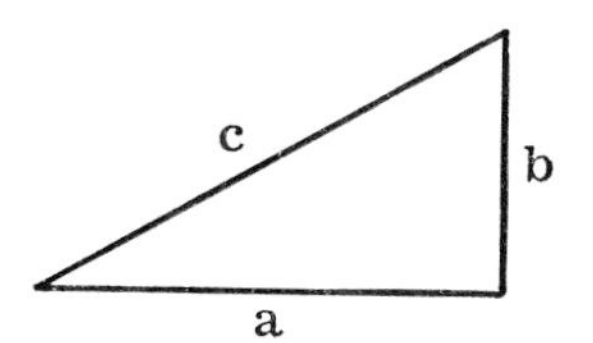

whose sides have an integral relationship to each other. Second, it refers to the fact that the Pythagoreans gave a formula for infinitely many such triangles. Namely, if m is odd and >1, set

$$a = m, \qquad b = \tfrac{1}{2}(m^2 - 1), \qquad \text{and} \qquad c = \tfrac{1}{2}(m^2 + 1) \tag{163}$$

Examples:

$$3^2 + 4^2 = 5^2$$

$$5^2 + 12^2 = 13^2$$

But there are also two senses in which this name, "Pythagorean" numbers, is seriously misleading. First, Neugebauer has shown that the Babylonians knew of the numbers of Eq. (162)—not *merely* those given by

Eq. (163)—at least 1,000 years before Pythagoras. Second, such a designation does not suggest, and indeed tends to conceal, the fact that *originally* the Pythagoreans thought that *every* right triangle would have its three sides in an integral relationship by a proper choice of the unit length. Furthermore, this belief was not a casual one but instead fundamental to the whole Pythagorean philosophy. When it was shattered by a number-theoretic discovery which the Pythagoreans made themselves, a profound crisis arose in this philosophy and in Greek mathematics.

Pythagoras (570?–500? B.C.) was born on the Greek island of Samos, traveled in Egypt, and perhaps in Babylonia, and founded a school and secret brotherhood in southern Italy. We need not go into the ethical doctrines that he expounded. On the scientific side, four subjects were studied; arithmetica (the theory of numbers), geometry, music, and spherics (mathematical astronomy). Of these four, arithmetica was considered the fundamental subject. In fact, the point of the Pythagorean philosophy was that Number is everything. We should make it clear at once that Number here means *positive integer*. There were no others. Since we are writing here on the theory of numbers, it behooves us to examine this far-reaching assertion in some detail.

The relationship between number and musical intervals was one of Pythagoras's first discoveries. If a stretched string of length, say, 12, sounds a certain note, the *tonic*, then it sounds the *octave* if the length is reduced to 6. It sounds the *fifth* (*do* to *sol*) if the length is reduced to 8, and the *fourth* (*do* to *fa*) if reduced to 9. So Harmony is Number. There follows a study of *means*. The fourth is the *arithmetic* mean of the tonic and octave, $9 = \frac{1}{2}(12 + 6)$, while the fifth is their *harmonic* mean, $\frac{1}{8} = \frac{1}{2}(\frac{1}{12} + \frac{1}{6})$, since its pitch is half-way between theirs. There also follows a study of *proportion*. The fifth is to the tonic as the octave is to the fourth, and the criterion of such proportionality is found in

$$8 \cdot 9 = 12 \cdot 6.$$

Since we may write this as

$$9 \cdot 8 = 12 \cdot 6,$$

we also have that the fourth is to the tonic as the octave is to the fifth, etc. The study of means and proportion was an important ingredient of Pythagoreanism.

The Pythagorean relationship between music and spherics is less convincing. The intervals between the seven "planets"—the Moon, the Sun, Venus, Mercury, Mars, Jupiter and Saturn—correspond to the seven intervals in the musical scale. This explains the Celestial Harmony, and shows that the Heavens too are essentially Number. We will see later how this mystic nonsense played a most important role in the history of science.

But the direct relation between number and spherics, without music as a middleman, was also known to Pythagoras from his travels in Egypt, and is worth more of our time. We shall not discuss Pythagorean astronomy in full. What we need to do is to understand a simple instrument called a *gnomon*, because it exemplifies the Pythagorean synthesis of spherics, geometry and arithmetica.

The gnomon is an L-shaped movable sundial used for scientific studies. It rests on one leg; the other is vertical. The length and direction of the shadow is measured at different times of the day and year. If the shadow falls directly on the horizontal leg at noon (when the shadow is shortest), that leg points north. The noon shadow changes length with the seasons—minimum at summer solstice and maximum at winter solstice. The sunrise shadow is perpendicular to the horizontal leg during the vernal or autumnal equinox. Thus the gnomon is a calendar, a compass and a clock. Pythagoras knew the world was a sphere—the gnomon measures latitude, it measures the obliquity of the ecliptic, etc. Here we have Solar Astronomy with Number (measurements) as the basis.

42. The Pythagorean Theorem

In all such shadow measurements the geometry of similar triangles and of right triangles is essential. A generation before Pythagoras, Thales of Miletus (a commercial center near Samos) also went to Egypt, studied mathematics, and started a school of philosophy. It is sometimes said that Pythagoras was one of his students. Plutarch tells the story that Thales determined the height of the Great Pyramid by comparing the length of the shadows cast by the Pyramid and by a vertical stick of known length. Some writers of mathematical history contest this, claiming that Thales did not know of the laws of similar triangles. We believe that he did, but we need not argue the point. It suffices for the argument which follows that the Pythagoreans did know about similar triangles, and this fact is not in question.

Nor do we raise the questions as to how and where Pythagoras "discovered" the Pythagorean Theorem. He may actually have learned of it from Egypt, for the "rope stretchers" there had long known how to construct right angles with a rope triangle of sides 3, 4 and 5; perhaps the Great Pyramid (2700 B.C.) had already been laid out in this way. But we *do* raise the question as to how Pythagoras *proved* (or thought he proved) the theorem, since this proof appears to be a critical step in the subsequent events.

We conjecture, on the basis of what we have already related, and upon subsequent events which we will relate presently, that the original proof ran as follows.

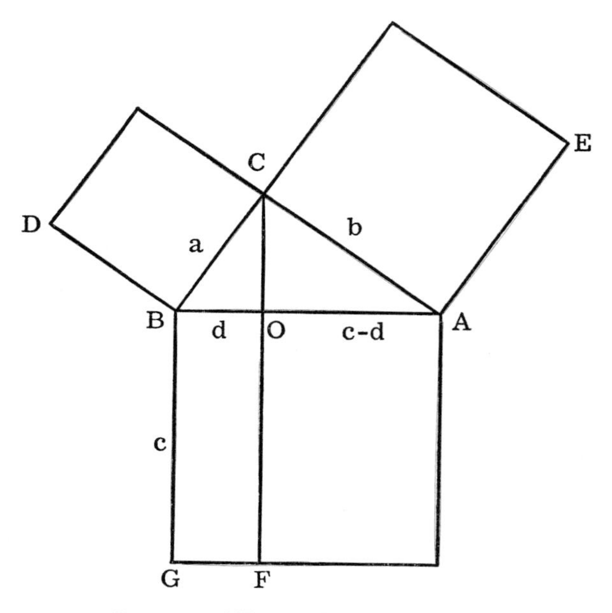

Draw the perpendicular COF. Find the greatest common measure of the four lines BC, CA, BO and OA. In terms of this length as a unit, let the four lines be of length a, b, d, and $c-d$ respectively. Since COB and ACB are both right angles and CBO equals itself, the triangles CBO and ABC are similar. Thus c is to a as a is to d. Here we have a third type of mean, a is the *geometric* mean of c and d, and

$$a^2 = cd.$$

Therefore the square CD equals the rectangle OG. Similarly CE equals AF, and the square on the hypothenuse equals the sum of the squares on the sides.

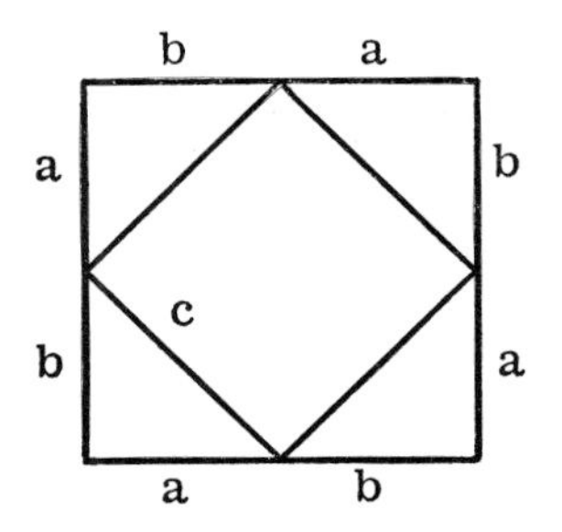

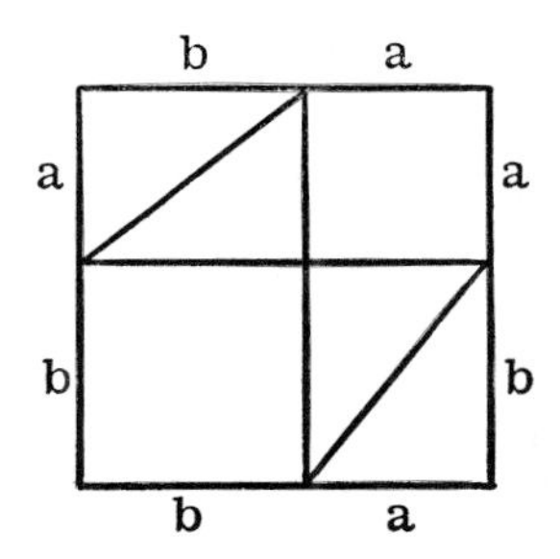

A number of historians have favored a different opinion—that Pythagoras's proof was a dissection proof such as that shown above. A square of side $a + b$ can be dissected into four triangles and the square c^2, or into four triangles and the two squares a^2 and b^2.

We think that this opinion is incorrect on three grounds.

(a) The suggested proof has *none* of the elements of Pythagoreanism—no proportion, no means, no "Number-as-Everything," no relation to spherics.

(b) The suggested proof is very clever, and appears to be of a sort that could be concocted *after* one knew the theorem to be true. But this implies a prior proof—or at least some serious evidence in the theorem's favor.

(c) The subsequent events, and their culmination in Euclid's *Elements*, are best explained in terms of the (fallacious) proof which we have suggested.

The Pythagorean derivation of Eq. (163) may date from the same (early) period as the Pythagorean Theorem. The names "square" number, "cube" number, "triangular" number, etc., all derive from the Pythagorean study of the relation between Number and form. The triangular numbers, 1, 3, 6, 10, etc., are the sums of consecutive numbers:

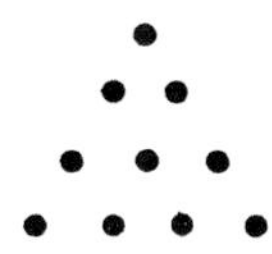

$10 = 1 + 2 + 3 + 4$, etc. The square numbers, 1, 4, 9, 16, etc., are the sums of successive *odd* numbers:

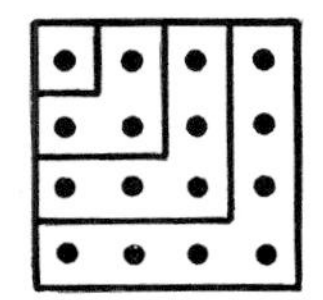

$16 = 1 + 3 + 5 + 7$, etc. The odd numbers the Pythagoreans called *gnomons*. It follows at once that if m is odd, and if m^2 is thought of as a gnomon of side $\frac{1}{2}(m^2 + 1)$, then

$$m^2 + [\tfrac{1}{2}(m^2 - 1)]^2 = [\tfrac{1}{2}(m^2 + 1)]^2.$$

This proves Eq. (163) "geometrically." And the first case of Eq. (163) is the Egyptian triangle, 3–4–5.

If we now look back at the illustration on page 123, we see the right triangular shadow and, framing the square on one side, the gnomon—which is really an odd number, etc. This was the Pythagorean synthesis at its best, and in its happy days—before the trouble began.

43. The $\sqrt{2}$ and the Crisis

The source of the trouble is attributed to Pythagoras himself. It is his

Theorem 56. *The equation*

$$2a^2 = c^2 \tag{164}$$

has no solution in positive integers.

Proof. Assume a solution with $(a, c) = g$. Let $a = Ag$ and $c = Cg$ and

$$(A, C) = 1. \tag{165}$$

Then

$$2A^2 = C^2.$$

But since C^2 is even, so must C be even. Let $C = 2D$ and

$$2A^2 = 4D^2, \quad \text{or} \quad A^2 = 2D^2.$$

Then A is also even, and since this contradicts Eq. (165), there is no solution.

This means that

$$\sqrt{2} \neq c/a.$$

It is not a ratio, therefore, from the modern point of view, it is an *irrational* "number." But an irrational number is no number at all—it is (via the Dedekind Cut) a class of classes of ordered pairs of numbers. It is totally "man-made," as L. Kronecker said, and thus is of dubious significance philosophically.

To the Pythagoreans, Theorem 56 was a terrible shock. It implies that in a 45° right triangle (with $b = a$), the hypothenuse and the side are *incommensurable*. There is *no common measure* such as we presumed in proof of the Pythagorean Theorem! The following serious consequences ensue.

(a) The proof is fallacious.
(b) The theorem is put in doubt.
(c) The theory of proportion, and of similar triangles, is put in doubt.
(d) The Pythagorean philosophy is largely undermined. For if Number (that is, positive integers), cannot even explain a 45° triangle, what becomes of the much more far-reaching claims?

The Pythagoreans were a secret society, and it is said that their discoveries were kept secret. But it is also said that Pythagoras's lectures were well-attended by the townspeople of Crotona. However contradictory this may appear, it is clear that Theorem 56 was highly embarrassing. The (unnamed) Pythagorean who first divulged this startling result is said to have suffered shipwreck in consequence, "for the unspeakable and invisible should always be kept secret."

At a later date a new embarrassment arose. While it was not of quite the same crucial character it may also have been considered important. The Pythagoreans knew of four regular polyhedra, and they associated these with the four "elements." The tetrahedron was fire, the cube was earth, the octahedron was air, and the icosahedron, water. But Hippasus, a member of the society, discovered the fifth regular polyhedron, the dodecahedron. By an ominous coincidence Hippasus, for divulging this discovery, was also shipwrecked and perished.

Far be it from us to suggest foul play on the basis of such flimsy evidence. Still, we recall that this was in southern Italy—the home of the Mafia—and that a cardinal principle of the Mafia is silence or quick retribution. The latter-day Mafia, in Chicago during the Prohibition era, was, as we know, involved in the numbers racket, and was also interested in fifths and fourths, and if squealers were seldom shipwrecked, they were often found, well-weighted, at the bottom of the Chicago river. Yet the parallel does not quite run true; it takes a rather vivid imagination to picture Little Caesar striding into the back room of the garage on Clark Street, and snarling, "OK, Louie, so you told about Gödel's Theorem! Now take dat!"

But returning to more solid ground, there is no questioning the fact that the problems raised by the $\sqrt{2}$ were most serious. We will examine the effects of this crisis upon geometry, "spherics," and arithmetica in the next three sections.

44. The Effect upon Geometry

If our supposition is correct, the order of the day at this point must have been to

(a) Devise a sound proof of the Pythagorean Theorem, and

(b) Devise a sound theory of proportion, which could handle incommensurate quantities, and therefore restore the important results concerning similar triangles.

Geometry as a deductive science probably began with the Pythagoreans. We see now that they had a strong motivation. When naive mathematics leads to paradoxes and contradictions, the day of rigorous mathematics begins. In the nineteenth century the paradoxes of the Fourier Series played a similar role in the motivation of rigorous mathematics; were it not too digressive, we should expound here on the parallelism of the problems created and of the answers found.

Instead, we skip over 200 years of Greek mathematics, and examine briefly the Greek answers to problems (a) and (b) above, as they appear in Euclid's *Elements*.

Euclid gives two proofs of the Pythagorean Theorem—in Book I, Prop. 47, and in Book VI, Prop. 31. Both proofs use (essentially) the same figure as we show on page 124. Neither proof has any relation whatsoever to the dissection figure on page 125. The first proof has nothing to do with similar triangles—these require a sound theory of proportion, and this is postponed to Book V. Book I is, so to speak, more elementary. It is clear, by reading it, that the main point of Book I is to prove the Pythagorean Theorem. This theorem is I, 47, and I, 48, the last proposition in Book I, is its converse. With few exceptions almost all of the previous theorems enter into the chain of proof leading to I, 47.

We show this in the following logical structure. The propositions labelled p are the "problems." We will discuss their role presently. The blank block under 46_p and 37 is inserted because both of these propositions depend upon both 31_p and 34.

The proof in I, 47 is based not on *similar* triangles, but on *congruent* triangles. Draw AD and CG in the figure on page 124. Then the triangles ABD and GBC are congruent. But the first equals half of the square CD, and the second, half of the rectangle OG. And so CD equals OG, etc.

The three theorems concerning congruent triangles—I, 4; I, 8; and I, 26; well-beloved of all high school geometry students—all play leading roles, as we see in the logical structure. The problems (bisect a line, an angle, construct a perpendicular, etc.) also play leading roles. Number plays no role. Proportion plays no role.

Book V gives the Eudoxus theory of proportion, the answer to problem (b), and in Book VI we find a second proof of the Pythagorean Theorem, similar to the one which we have attributed to Pythagoras—but now based upon the logically sound Eudoxus theory. There can be no doubt that Euclid knew of the earlier "proof," and also what was wrong with it.

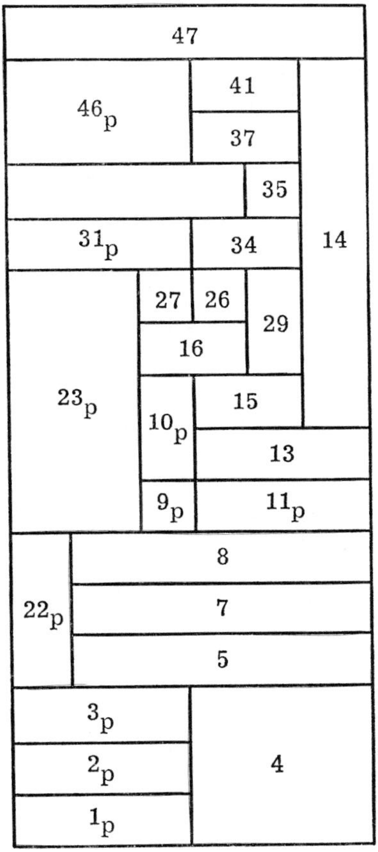

In conclusion we would point out that three important "peculiar" aspects of the *Elements* all bear testimony to the original Pythagorean "proof" and to the subsequent crisis over the $\sqrt{2}$.

(a) In elementary teaching the "problems" are often thought of as exercises, or as applications. Euclid has no use whatsoever for exercises or applications. The problems are *proof* that any construction called for in the proof of a theorem is indeed possible. The original mistake of Pythagoras, "Find the greatest common measure, etc.," was not to be repeated.

(b) Number is expelled from Geometry. Much nonsense has been written on this point. It has been called a peculiarity of the Greek "mind"—a preference for form rather than number—a greater ability in geometry than arithmetic, etc. There is no basis for this. Euclid has three books on the theory of numbers. The origins of Greek mathematics in Egypt and Babylonia were definitely numerical. Pythagoras's opinion of Number we know. The expulsion of number from geometry was solely due to the problems raised by the $\sqrt{2}$.

(c) Euclid's proof of I, 47 is seldom appreciated in its historical context. No doubt Euclid "liked" the logical simplicity of the fallacious Pythagorean proof. But to postpone a proof of the Pythagorean Theorem until after the "advanced" Eudoxus theory can be studied is undesirable. Therefore Euclid gives the most elementary proof he can find, while keeping as close as possible to the original Pythagorean structural framework. When Schopenhauer criticized this Euclid proof of I, 47 as a "mousetrap proof," "a proof walking on stilts," etc., he showed that he had little appreciation of the historical, mathematical, and even philosophical points which were involved.

45. The Case for Pythagoreanism

The most important problem concerning the integers is the determination of their role in Nature. The Pythagoreans said Number is everything, but, aside from the analysis of music, we cannot say that they made a good case for this assertion. Nor could they be expected to do so, with science at such a primitive level. The mystic and numerological aspects of Pythagoreanism we now regard most unfavorably. However, these aspects can be ignored. The real difficulty with Pythagoreanism stems from the $\sqrt{2}$ and its corollary that in the analysis of *continuous* magnitude the integers (as such) do not quite suffice.

If we ask whether modern physical scientists believe that the world can be best understood numerically, the answer is yes—practically all of them do. But here "numbers" are no longer confined to integers; they also include real numbers, vectors, complex numbers, and other generalizations. The founders of modern physical science (Galileo, Kepler, and others) did not have a rigorous theory of real numbers, but they had the practical equivalent, namely, decimal fractions. These, of course, the Greeks did not have. The formulation of the laws of nature in terms of ordinary differential equations (Newton), and in terms of partial differential equations (Euler, D'Alembert, Fourier, Cauchy, Maxwell), appeared to further weaken the role of integers in Nature and to strengthen that of real numbers. But even here we may note that while the variables in an equation are continuous, the order of the equation, and the *number* of variables in it, are integers—a point that should not be neglected.

A philosophy which interprets the world numerically, in the general sense of real numbers, we may call *New Pythagoreanism*, whereas one that insists that the integers are fundamental—not only mathematically, but also physically—we call *Old Pythagoreanism*. We now inquire whether a case can be made for Old Pythagoreanism. To determine this we must examine a list of some of the key discoveries in physical science.

(a) Galileo (1590) found that during successive seconds from the time at which it starts falling, a body falls through distances proportional to 1, 3, 5, 7, etc., so that the total distance fallen is proportional to the square of the time. Here we have square numbers arising as sums of the odd numbers (gnomons!).

(b) Johannes Kepler was an out-and-out Pythagorean*—one who really believed in the Harmony of the Spheres (page 122), etc. He sought for many years to find accurate numerical laws for astronomy expressing such "harmonies" and in 1619 he discovered his important Third Law—the squares of the periods of the planets are proportional to the cubes of their mean distances from the sun.

(c) Even before Newton's *Principia* (1687) it was known to Robert Hooke, Christopher Wren, and others, that the integer exponents in (a) and (b) imply that each planet has an acceleration toward the sun which is inversely proportional to the square of its distance from the sun.

(d) Inspired by Newton's Law of Gravity (c), Charles Coulomb (1785) determined, with a torsion balance, that electrostatic forces were also inverse square. Henry Cavendish (1773, unpublished) had already obtained the same law by another method—one which is most instructive for our present discussion. The experiment was repeated by Maxwell a hundred years later. The experimenter enters a large hollow electrical conductor. The conductor is charged to a high potential and the experimenter attempts to measure a change of potential on the *inside* surface. He finds nothing—within experimental error. In this way Maxwell established that the exponent is -2 with a probable error of $\pm 1/21600$. Later experiments reduced the possible deviation from -2 even further.

The point involved is this: a New Pythagorean might say that Coulomb's results merely indicate that the exponent is *approximately* equal to -2. But the Cavendish-Maxwell experiment not only suggests that it is *exactly* -2 but also suggests the "reason" for this. Mathematically the only law of force which would behave in this way is one whose *divergence* is zero—that is, one that falls off radially in such a way as to just compensate for the increase in the area of a spherical shell with its radius. Now this area increases with the square of the radius, and this is so because we live in a space of three dimensions. In effect, then, the fact that the exponent -2 is an integer is directly associated with the fact that the dimensionality of space is an integer.

(e) From this interpretation of Coulomb's Law (the divergence is zero), from a similar, inverse square, electromagnetic law due to André Marie Ampère (1822), and from other experimental results, James Clerk Maxwell was led to the electromagnetic wave equations in 1865. While the dependent

* He even suggested the possibility that the soul of Pythagoras may have migrated into his own.

and the independent variables here are both continuous, we shall see that in some respects the number of independent variables, 4, and of dependent variables, 6, is more fundamental. We will pick up this thread presently.

(f) Proust's Law of Definite Proportions (1799) and Dalton's Law of Multiple Proportions (1808) in chemistry directly imply an Atomic Theory of matter. The integral ratios in the second law exclude any other interpretation. Further, it appears that chemical affinity involves integers directly, namely the valence of the elements.

(g) In exact analogy, Lisle's Law of Constant Angles (1772) and Haüy's Law of Rational Indices (1784) for crystals directly imply that a crystal consists of an integral number of layers of atoms. Again, the integral ratios in the second law exclude any other interpretation. Further, there is a direct relationship between number and form, e.g., the six-sided symmetry of frozen H_2O.

(h) The ratio of the two specific heats of air is 7/5 and of helium is 5/3. While the (New Pythagorean) phenomenological theory (thermodynamics) cannot explain these integral ratios at all, the atomic theory (f) explains them easily (Boltzmann). By a similar argument Boltzmann explains the Dulong-Petit Law for the specific heats of solids.

(i) Faraday's Law of Electrolysis (1834) states that the weight of the chemical deposited during electrolysis is proportional to the current and time. If chemical weight is atomic, from (f), then this law implies that electricity is also atomic. Such electric particles were called electrons by Stoney (1891). We will pick up this thread presently.

(j) In 1814 Joseph von Fraunhofer invented the diffraction grating. A glass plate is scratched with a large number of parallel, uniformly spaced, fine lines. This integral spacing produces an optical spectrum, since parallel light of a given wavelength, shining through the successive intervals on the glass, will be diffracted only into those directions where the successive beams have path lengths that differ by an integral number of wavelengths.

(k) The simplest spectrum is that of hydrogen. The wavelengths of its lines have been accurately determined, (j). In 1885 Balmer found that these wavelengths are expressible by a simple formula involving integers.

(l) Pieter Zeeman (1896) discovered that the lines of a spectrum are altered by a magnetic field, and H. A. Lorentz at once devised an appropriate theory. The radiating atoms (f) contain electrons (i) whose oscillations produce the spectrum by electromagnetic radiation (e). The frequency of the oscillations (and therefore also their wavelength) is changed by the action of the magnetic field upon the electrons.

(m) From Maxwell's Equations (e) and thermodynamics, Ludwig Boltzmann (1884) derived Stefan's Law of Radiation (1879). This states that a blackbody radiates energy at a rate proportional to the fourth power of its absolute temperature. We note that although electromagnetism and thermodynamics are both theories of continua (New Pythagoreanism) the real point of the law is the *exponent*. Here again the exponent 4 is said to be exact and, in fact, even a casual examination of Boltzmann's derivation shows that this exponent equals the number of independent variables in the wave equation—the three of space and one of time. Just as $2 = 3 - 1$ in (d), so does $4 = 3 + 1$ here.

(n) But this Boltzmann theory involving continua (m) and his other theory (h) involving atoms contradict experiment when combined theoretically. Thus if electromagnetic radiation is produced by oscillating electrons (l), the statistical theory of equilibrium which Boltzmann developed for (h) does not imply Stefan's Radiation Law (m). It implies instead the so-called Rayleigh-Jeans Law, which does not agree with experiment, and in fact asserts that an infinite amount of energy will be radiated! In plain language this erroneous law implies that equilibrium is not possible at temperatures above absolute zero.

To save the situation, that is to preserve both Stefan's (m) and Lorentz's (l), Max Planck (1900) found it necessary to assume that the energy was not radiated continuously but discretely in *quanta*. He gives

$$E = h\nu$$

where E is the energy of the quantum, ν is its frequency, and h is a constant. It is interesting to note that this Planck constant h enters into a related radiation law (Wien's Displacement Law) in the form of a *ratio*, k/h, where k is the Boltzmann constant. Just as h is a measure of the energy per *quantum*, so k is a measure of the energy per *atom*. The ratio k/h is determined experimentally. If atoms are "small," then so are quanta "small," but if matter is not continuous—that is, if $k > 0$—then neither is energy continuous, since $h > 0$.

But Planck was a New Pythagorean and did not like his (discrete) quanta. He sought for years to circumvent his own (fundamental) discovery. But the logic is clear. Just as discrete matter implies discrete electricity in (i) so does discrete matter imply discrete energy here—for the *ratio* may be determined experimentally in either case.

(o) Einstein accepted quanta "heuristically" and in 1905 he used them to explain photoelectricity.

(p) In the same year, but in quite a different vein, he also developed relativity. The Michelson-Morley experiment (1887) had suggested that Maxwell's Equations (e) must remain invariant to observers traveling with different velocities. The consequences of such an assumption are that time and space are no longer absolute and distinct, but are related by the Lorentz Transformation. In the hands of H. Minkowski (1908) this led to the four-dimensional space-time continuum. In this theory particular importance is attached to vectors with four components. One such vector is a space-time displacement. Another, which we will need soon, is the momentum-energy vector, three components of momentum and one of energy. A skew-symmetric tensor in this four-dimensional world has six components—four things taken two at a time. The most important example is the electromagnetic field—three components of electric field, and three of magnetic.

(q) This recalls the fact that the Pythagoreans also considered four to be especially important. Thus (they say), the soul is related to fire, and fire, as we indicated before, is a tetrahedron, and a tetrahedron has both four vertices and four faces, and is the smallest regular polyhedron. The reader may well consider that we should hastily stow this back in the closet—and lock the door. But we have our purpose, and since we have raised the point let us examine it for a moment.

The Pythagoreans say that a point is of no dimension, two points form a line, three points a surface, and four a solid. A tetrahedron has two special properties: it is the smallest polyhedron, and it has the same number of vertices and faces (i.e., it is self-dual). Both properties follow from the fact that its number of vertices is one more than the dimensionality of space. Let us admit, then, that four is important to Pythagoras for the same simple reason that it is important to Einstein, Minkowski, Stefan and Boltzmann: $4 = 3 + 1$.

But why should fire be a tetrahedron? The reader knows that the spectacular part of fire is the radiant heat and light, and that this is electromagnetic, and that the six components of this field are obtained by taking the four dimensions of space-time two at a time (p). So likewise the six edges of the tetrahedron join the four vertices two at a time, and also are the intersections of the four faces two at a time. But we do not insist upon it. If the reader can find a more fitting regular polyhedron for fire let him do so. We now close the closet and return to experimental facts.

(r) A most important discovery, and one which is very instructive for our present investigation since it combines New and Old Pythagoreanism, is Mendeléeff's Periodic Table of the chemical elements (1869). If the elements are listed in order of their atomic weights (f), then chemical, spectroscopic, and some other physical similarities recur periodically. But there were many imperfections and many questions arose. Tellurium weighs more than iodine. But if placed in the table in that order these elements clearly fall into the wrong groups. Again, the position of the rare earths and the numerous radioactive decay products was not clear. The rare gases were entirely unanticipated. Further, the table is not strictly periodic but has periods of length 2, 8, 18, and 32. Why these periods should all be of the form $2n^2$ was not clear. Indeed, how could it be—for what can mere *weight* have to do with these other properties?

(s) In 1911 C. G. Barkla found, by x-ray scattering, that an atom contains a number of electrons approximately equal to one-half of its atomic weight.

In the same year E. Rutherford found, by alpha particle scattering, that the (compensating) positive charge, and with it most of the mass, was concentrated at the center of the atom. This positive charge was about one-half of the atomic weight. There followed Rutherford's theory of the atom—a miniature "solar" system with the light, negatively charged electrons bound to the heavy, positively charged nucleus by inverse square Coulomb forces (d).

(t) In 1913 Niels Bohr assumed that the hydrogen atom had this (simplest) Rutherford structure (s)—one proton as a nucleus and one electron as a satellite. With the use of Planck's $E = h\nu$, (n), he deduced the Balmer formula (k) with great precision. However, he had to assume that the electron could have a stable orbit only if its angular momentum were an integral multiple of $h/2\pi$. That is,

$$mvr = nh/2\pi$$

with m the electron's mass, r the orbit's radius, v the electron's velocity, and h Planck's constant. The integer n, the *principal quantum number*, made no sense in the New Pythagorean theories then in vogue, but its acceptance was forced by the remarkable accuracy of the theory's predictions.*

(u) 1913 was a good year for Old Pythagoreanism. Soddy and Fajans found that after radioactive emission of an alpha particle (charge $+2$) the resulting element is two places to the left in the periodic table, whereas emission of a beta particle (charge -1) results in a daughter element one place to the right. Together with the earlier results in (s) this Displacement Law makes it clear that *atomic number*, not atomic weight, is the important factor. This integer is the positive charge on a nucleus, the equal number of electrons in that atom, and the true place in the table of elements. This was explicitly stated by van den Broek and rapid confirmation was obtained by Moseley (w).

(v) In 1912 von Laue made the very fruitful suggestion that a crystal (g) would act like a diffraction grating (j) for radiation of a very short wavelength.

---

* While Niels Bohr was applying numbers to the analysis of spectra, his brother, Harald Bohr, was applying a generalized spectral analysis (almost periodic functions) to the analysis of number (prime number theory).

(w) Henry Moseley (1913) used von Laue's suggestion (v) to measure the (very short) wavelengths of x-rays. Optical spectra, like chemical behavior, are due to the outer electrons in an atom, and thus have a periodic character. But x-ray spectra are due to the inner electrons, and these electrons are influenced almost solely by the charge on the nucleus. Moseley's photographs show a most striking monotonic variation of the x-ray wavelengths with atomic number.

Atomic *number* at once cleared up most of the difficulties in (r). But what about $2n^2$?

We note in passing a remarkable neck-and-neck race of x-rays and radioactive radiation:

	X-Rays		*Radioactivity*	
Discovery	1895	Roentgen	1896	Becquerel
Atomic Structure	1911	Barkla	1911	Rutherford
Atomic Number	1913	Moseley	1913	Soddy

(x) In 1923 L. de Broglie applied relativistic invariance of four-vectors (p) to Planck's $E = h\nu$, (n). The energy E and the time associated with the frequency ν are merely single components of two four-vectors. The remaining three components of momentum and of space, respectively, (p), must be similarly related. Thus a particle of momentum mv should have a (de Broglie) wavelength λ given by

$$\lambda = \frac{h}{mv}.$$

When this is applied to Bohr's

$$mvr = nh/2\pi$$

one obtains

$$n\lambda = 2\pi r.$$

Thus the matter wave has exactly n periods around the circumference of the orbit and the interpretation of the electron's stability is that it constitutes a standing wave.

(y) This conception was refined in the Schroedinger Wave Equation (1926). Here there are *three* quantum numbers n, l, and m corresponding to the dimensionality of space. In polar coordinates the wave functions corresponding to l and m are spherical harmonics—no, not Harmony of the Spheres—but very close to it. It further develops that the integer l can equal $0, 1, 2, \cdots, n-1$ while m can equal $-l, -l+1, \cdots, l-1, l$.

For $n = 4$, for example, we have 16 possible states:

values of m

3	2	1	0
2	1	0	−1
1	0	−1	−2
0	−1	−2	−3
↑	↑	↑	↑
$l =$ 0	1	2	3

Gnomons!

(z) But a fourth quantum number was already waiting. In 1925 Uhlenbeck and Goudsmit discovered the spin of the electron. This gives rise to a fourth number s which can take on *two* possible values. When this fourth coordinate is added, with its astonishing rounding out of the little "solar" system by rotating the "planets" and thus simulating time, we obtain the $2n^2$ states which correlate with the periods in the periodic table. But we must distinguish—and also associate—two different "harmonies" here. In one atom an electron can go from state to state; thus giving rise to the spectrum. This is the first "harmony." On the other hand, as we go through the periodic table, adding one new electron each time, the new electrons will also take on these distinct quantum states according to the Pauli Exclusion Principle (1925). This gives rise to the periodic table—the second "harmony."

Before the rare gases were discovered it seemed as though the (lighter) elements in the periodic table had a period of 7, not 8, and Newland (1864) called this the Law of Octaves. He was an Old Pythagorean, but he lacked the facts.

If we thought it necessary to strengthen the case we could continue and discuss isotopes (Soddy); $hc/2\pi e^2 = 137$ (Eddington); "magic" numbers (Mayer); "strangeness" numbers (Gell-Mann); etc. It is not a coincidence, for example, that the three nuclei which are fissionable with slow neutrons, U^{233}, U^{235}, and Pu^{239}, all contain an even number of protons and an odd number of neutrons.

However it is not our purpose to write a history of science. We asked whether there is a case for Old Pythagoreanism. We conclude that there is —and a strong one. Henceforth we shall call it Pythagoreanism.

Exercise 104. Draw a diagram showing the historical-logical structure of the discoveries (a) to (z) discussed above.

46. Three Greek Problems

We now return to number theory and consider three problems which are immediately suggested by (the troublesome) Theorem 56. We recall that this theorem stated that the equation

$$c^2 = 2a^2 \tag{166}$$

has no solution in positive integers. The *first* problem is that of generalizing this theorem. While the $\sqrt{2}$ is encountered in a 45° right triangle (one-half of a square), the $\sqrt{3}$ is similarly encountered in a 30°–60° right triangle (one-half of an equilateral triangle), and the corresponding equation is

$$c^2 = 3a^2. \tag{167}$$

Equation (167) again has no solution in positive integers, or, we may say, the $\sqrt{3}$ is irrational.

Plato states that Theodorus the Pythagorean (ca. 400 B.C.) showed that $\sqrt{3}$, $\sqrt{5}$, $\sqrt{6}$, $\sqrt{7}$, $\sqrt{8}$, $\sqrt{10}$, $\sqrt{11}$, $\sqrt{12}$, $\sqrt{13}$, $\sqrt{14}$, $\sqrt{15}$, and $\sqrt{17}$ were all irrational, "beyond which for some reason he did not go." The implication is that Theodorus had no *general* approach to the problem. With the use of the unique factorization in Theorem 7, however, it is very easy to prove the more general

Theorem 57. *The equation*

$$c^n = Na^n \tag{168}$$

has no solution in positive integers unless N is the nth power of an integer.

Proof. If c and a are written in standard form:

$$c = p_1^{\alpha_1} p_2^{\alpha_2} \cdots, \qquad a = q_1^{\beta_1} q_2^{\beta_2} \cdots$$

we see that c^n, a^n, and thus also c^n/a^n, have all the exponents in their standard factorizations divisible by n. Therefore $N = c^n/a^n$ is an nth power.

There are many deeper solved and unsolved problems concerning irrational and transcendental numbers, but it would be digressive to discuss them now.

The *second* problem arises by modifying Eq. (166) to read

$$c^2 - 2a^2 = \pm 1. \tag{169}$$

The motivation is clear. The right side of Eq. (169) cannot be replaced by zero. To best approximate an isosceles right triangle we seek sides a, and "diagonal" c, with the right side of Eq. (169) having the *smallest* magnitude possible. The corresponding isosceles triangle approximates a right triangle and the ratio c/a is a rational approximation of the $\sqrt{2}$.

It is interesting that the opposite strategy leads to (essentially) the same problem. Let the triangle be a *right* triangle whose (integral) sides differ by as little as possible, that is, let $\bar{b} = \bar{a} + 1$ in Eq. (162). Then from $\bar{a}^2 + \bar{b}^2 - \bar{c}^2 = 0$ we have

$$2\bar{a}^2 + 2\bar{a} + 1 - \bar{c}^2 = 0 \quad \text{or}$$
$$(2\bar{a} + 1)^2 - 2\bar{c}^2 = -1. \tag{170}$$

We therefore require a solution of Eq. (169) with c odd $(= 2\bar{a} + 1)$, $a = \bar{c}$, and -1 on the right.

The Pythagoreans knew at least some of the solutions of Eq. (169). But Theon of Smyrna (ca. A.D. 130) gave

Theorem 58. *Let the "side" and "diagonal" numbers a_n and c_n be defined by*

$$a_1 = 1, \qquad c_1 = 1;$$
$$a_2 = 2, \qquad c_2 = 3;$$
$$a_3 = 5, \qquad c_3 = 7;$$

and, in general,

$$a_{n+1} = a_n + c_n\,, \qquad c_{n+1} = 2a_n + c_n. \tag{171}$$

Then

$$c_n^2 - 2a_n^2 = (-1)^n. \tag{172}$$

Proof. From Eq. (171)

$$\begin{aligned} c_{n+1}^2 - 2a_{n+1}^2 &= (2a_n + c_n)^2 - 2(a_n + c_n)^2 \\ &= 2a_n^2 - c_n^2 \\ &= -(c_n^2 - 2a_n^2). \end{aligned}$$

Since $c_1^2 - 2a_1^2 = -1$, Eq. (172) follows by induction.

Several comments are in order. Equation (171), in fact, gives *all* the solutions of Eq. (169), but that has not yet been demonstrated. The source of the solution Eq. (171) is not indicated here but will be revealed later. Finally we note that the right triangles obtained by Eq. (170), from the side and diagonal numbers for n odd (and >1), are given by the triples:

$$(3, 4, 5)\,;\ (20, 21, 29)\,;\ (119, 120, 169)\,;\ \text{etc.}$$

These agree with the Pythagorean sequence, Eq. (163):

$$(3, 4, 5)\,;\ (5, 12, 13)\,;\ (7, 24, 25)\,;\ \text{etc.},$$

only in the first triangle.

Theorem 58 has an important generalization but some modification is necessary. For example, if we replace 2 by 3 in Eq. (169) and choose the negative sign:

$$c^2 - 3a^2 = -1 \tag{173}$$

we obtain an equation with no solution. That is clear since it implies $c^2 \equiv -1 \pmod 3$, and we know that is impossible. But if we choose the plus sign and if N is not a square, the equation

$$c^2 - Na^2 = 1 \tag{174}$$

always has infinitely many solutions. This important theorem of Fermat we postpone until later.

If $N = 3$, we have

$$c^2 - 3a^2 = 1 \tag{175}$$

and, while Eq. (173) is impossible,

$$c^2 - 3a^2 = -2 \tag{176}$$

is not. In his famous *Measurement of a Circle* Archimedes obtains

$$3\tfrac{1}{7} > \pi > 3\tfrac{10}{71}$$

and in deducing these inequalities he uses

$$\frac{1351}{780} > \sqrt{3} > \frac{265}{153}.$$

The reader may verify that these good approximations to $\sqrt{3}$, (call them c/a), satisfy Eqs. (175) and (176) respectively, so that Archimedes knew at least some solutions of these equations.

Exercise 105. From one of these Archimedean approximations to the $\sqrt{3}$, and by an approach similar to Eq. (170), deduce the fact that (451, 780, 901) gives a right triangle which is approximately 30°–60°.

The last exercise, and the two series of Pythagorean numbers given above, suggest the *third* problem—that of finding *all* solutions of

$$a^2 + b^2 = c^2.$$

This is solved by

Theorem 59. *If a, b, and c are positive integers which satisfy*

$$a^2 + b^2 = c^2, \tag{177}$$

it is sufficient if they are given by

$$a = s(2uv), \qquad b = s(u^2 - v^2), \qquad \textit{and} \qquad c = s(u^2 + v^2), \tag{178}$$

with $u > v$, and Eq. (178) *is also necessary providing we are willing to interchange the formulae for a and b if this is necessary.*

COMMENT: The sufficiency was given by Euclid, Book X, Prop. 28, 29, but was known to the Babylonians more than 1,000 years earlier (see page 121).

PROOF. Since

$$(2uv)^2 + (u^2 - v^2)^2 = (u^2 + v^2)^2$$

is an identity, the sufficiency of Eq. (178) is obvious. Suppose $(a, b) = s$ in Eq. (177). Then $s|c$ and let $a = sA$, $b = sB$, and $c = sC$. Then

$$A^2 + B^2 = C^2 \tag{179}$$

with A, B, and C all prime to each other. A and B are not both odd, for if so $A^2 + B^2 = C^2$ is of the form $4m + 2$, and this is impossible. Nor are they both even, since $(A, B) = 1$. Without loss of generality let A be even and B be odd, and therefore C is odd. Then

$$\left(\frac{A}{2}\right)^2 = \frac{C - B}{2} \cdot \frac{C + B}{2}. \tag{180}$$

But $(C + B)/2$ and $(C - B)/2$ are prime to each other, for, if not, their sum C and difference B would not be either. By Theorem 7 and Eq. (180), $(C + B)/2$ and $(C - B)/2$ are therefore squares, say u^2 and v^2.

Therefore

$$A = 2uv, \qquad B = u^2 - v^2, \qquad \text{and} \qquad C = u^2 + v^2. \tag{181}$$

Then Eq. (178) follows.

Corollary. *All Pythagorean numbers*

$$A^2 + B^2 = C^2$$

with A, B, and C prime to each other, and with A even, are given by Eq. (181) *where u and v are prime to each other, one being odd and one even. These triples are called* primitive *triangles.*

The 12 smallest primitive triangles listed according to hypothenuse are:

A	B	C	u	v	$\mid A - B \mid$
4	3	5	2	1	1
12	5	13	3	2	7
8	15	17	4	1	7
24	7	25	4	3	17
20	21	29	5	2	1
12	35	37	6	1	23
40	9	41	5	4	31
28	45	53	7	2	17
60	11	61	6	5	49
56	33	65	7	4	23
16	63	65	8	1	47
48	55	73	8	3	7

Exercise 106. In how many primitive triangles is 85 the hypothenuse? What about 145?

Exercise 107. If u_0 and v_0 are prime to each other and *both* odd, show that the A, B, C obtained from Eq. (181) equal $2B'$, $2A'$, $2C'$ for some primitive triangle: A', B', C'. Determine the u and v for this triangle in terms of u_0 and v_0.

47. Three Theorems of Fermat

Just as Theorem 56 led the Greeks to the three problems discussed above, so did Theorem 59 lead Fermat to three important theorems. Each of these, in turn, led to an important *branch* of number theory. We will prove none of these theorems in this section but will state all three—in a survey fashion.

Perhaps the most important is

Theorem 60. *Every prime of the form* $4m + 1$ *is the sum of two squares in a unique way.*

Examples:

$$5 = 1^2 + 2^2, \qquad 17 = 1^2 + 4^2, \qquad 37 = 1^2 + 6^2,$$

$$13 = 2^2 + 3^2, \qquad 29 = 2^2 + 5^2, \qquad 41 = 4^2 + 5^2.$$

This theorem, which had already been *stated* by Girard several years earlier, is, of course, suggested by the third column of the foregoing table and the formula $C = u^2 + v^2$. In example (h) of Theorem 48, page 106, we have seen that if C is prime it is of the form $4m + 1$. But to prove Theorem 60 we need the (harder) converse and also the uniqueness. The theorem is rather surprising, since primality is a purely multiplicative concept. What can primality have to do with a *sum* of squares? We will return to this point and theorem.

With Eqs. (170) and (171) we obtained infinitely many primitives with $|A - B| = 1$. The column $|A - B|$ above has familiar looking numbers, from our studies of the factors of M_p, and suggested to Frenicle and Fermat that every prime of the form $8m \pm 1$ is the difference of the legs of infinitely many primitive triangles. Since

$$|A - B| = |(u - v)^2 - 2v^2|$$

the implication is that every prime p of the form $8m \pm 1$ can be written as

$$\pm p = x^2 - 2y^2$$

in infinitely many ways. Together with Theorem 60 we are led to consider the numbers

$$x^2 + Ny^2$$

for every integer N. This brings us to an extensive subject—that of *binary quadratic forms*. We may note that while perfect numbers and periodic decimals lead to quadratic residues only at a deeper level, with Pythagorean numbers they arise at once. For if a prime p is given by

$$p = x^2 + Ny^2,$$

then, since $(y, p) = 1$, we have

$$(y^{-1}x) \equiv \sqrt{-N} \pmod{p}.$$

It is interesting to note that the two square roots which were most fruitful historically in forcing an extension of the number system, namely $\sqrt{-1}$ and $\sqrt{2}$, were also those which arose earliest in these binary quadratic forms, $x^2 + Ny^2$.

Further examination of the column C raises other questions. The hypothenuse 65 arises twice:

$$65 = 7^2 + 4^2 = 8^2 + 1^2.$$

In how many ways is an integer a sum of two squares?

And, of course, some numbers *cannot* be written as sum of two squares. But, of these, some are a sum of three squares, and some of four. Thus

$$14 = 9 + 4 + 1; \qquad 7 = 4 + 1 + 1 + 1.$$

Following an earlier *statement* by Bachet, Fermat proved

Theorem 61. *Every positive integer n is expressible as*

$$n = w^2 + x^2 + y^2 + z^2 \tag{182}$$

where w, x, y, and z are integers, positive or zero.

Like Theorem 60, Theorem 61 is rather surprising, and rather hard to prove. Euler was unable to prove it although he worked on it for years. Through its generalization, *Waring's Problem*, it became a major source of *additive number theory*. A sketch of a proof of Theorem 61 is given in Exercises 31S–33S, on page 209. The first published proof is due to Lagrange.

From a sense of symmetry the reader probably can guess what comes next. If a sum of two squares leads us to consider a sum of four squares on the one hand, it should also lead us to consider a sum of two fourth powers on the other. In the foregoing table *either* A (as in 4, 3, 5) or B (as in 40, 9, 41) may itself be a square, but not both *simultaneously*. This result is closely related to an impossible problem of Bachet—to find a Pythagorean triangle whose *area* is also a square. (We may call this the problem of "squaring the triangle"—in integers.) This problem may be shown to imply the following condition:

$$a^4 - b^4 = c^2. \qquad (183)$$

Fermat proved Eq. (183) impossible, and similarly he proved

Theorem 62. *The equation*

$$a^4 + b^4 = c^2 \qquad (184)$$

has no solution in positive integers.

Corollary. *The equation*

$$a^4 + b^4 = c^4 \qquad (185)$$

has no solution in positive integers.

PROOF OF THE COROLLARY. A fourth power is a square.

We will prove Theorem 62 later. The corollary is in striking contrast with Theorem 59 where there are infinitely many solutions. The corollary is the only easy case of Fermat's Last "Theorem." We will consider this celebrated conjecture in the next section. While it is sometimes stated to be an isolated problem—of no special significance—it was, in fact, one of the main sources of *algebraic number theory*.

48. FERMAT'S LAST "THEOREM"

It is well known that Fermat wrote that he had "a remarkable proof" of

Conjecture 16. *The equation*

$$a^n + b^n = c^n \qquad (186)$$

has no solution in positive integers if $n > 2$.

The Corollary of Theorem 62 is, of course, the case $n = 4$. The reader probably knows that no *general* proof has been found, although "it has been attempted by Euler, Legendre, Gauss, Abel, Dirichlet, Cauchy, Kummer," etc.; that Paul Wolfskehl, a wealthy German interested in number theory, offered a reward of 100,000 marks in 1908; that Hugo Stinnes, a wealthy German not interested in number theory, helped bring on the German Inflation in the 1920's and thus (incidentally) reduce the value of this prize considerably; and that (nonetheless) much further effort has been expended by thousands of professionals and amateurs with no conclusive result. According to Professor Mordell, there are easier ways to make money than by proving Fermat's Last Theorem.

We will first give an interesting approach which makes the conjecture plausible. The reader knows that if $g(x)$ is a rational function of x,

$$\int g(x)\,dx$$

is integrable in terms of *elementary* functions—that is, a finite combination of algebraic, trigonometric and exponential functions together with their inverses. Or, again, say,

$$\int \sqrt{1 - x^2}\,dx \tag{187}$$

is so integrable. But

$$\int \sqrt{1 - x^4}\,dx \tag{188}$$

is not elementary—it is an elliptic integral.

Chebyshev has proven that if U, V, and W are rational numbers, then

$$\int x^U (A + Bx^V)^W\,dx \tag{189}$$

is integrable in terms of elementary functions if and only if

$$\frac{U+1}{V}, \qquad \text{or} \qquad W, \qquad \text{or} \qquad \frac{U+1}{V} + W \tag{190}$$

is an integer. In Eq. (188) we have $A = -B = 1$, $U = 0$, $V = 4$, and $W = \frac{1}{2}$. But neither $\frac{1}{4}$, nor $\frac{1}{2}$, nor $\frac{3}{4}$ is an integer.

If in Theorem 59 we set $x = a/c$ and $y = b/c$ and $t = v/u$ we find that all *rational* solutions x and y of

$$x^2 + y^2 = 1$$

are given by

$$x = \frac{2t}{1 + t^2} \qquad \text{and} \qquad y = \frac{1 - t^2}{1 + t^2} \tag{191}$$

where t is an arbitrary rational number.

Now, in Eq. (186), let us similarly write $x = a/c$ and $y = b/c$ and generalize the exponent n to be any rational number k. Thus

$$x^k + y^k = 1. \tag{192}$$

We now ask, following the example of Eq. (191): Are there rational functions

$$x = f(t) \qquad \text{and} \qquad y = g(t)$$

such that Eq. (192) is satisfied identically?

If $k = 1/q$, for q a nonzero integer, the answer is yes, since we may set

$$x = t^q \qquad \text{and} \qquad y = (1 - t)^q.$$

And, if $k = 2/q$ the answer is yes, since we may set

$$x = \left(\frac{2t}{1 + t^2}\right)^q \qquad \text{and} \qquad y = \left(\frac{1 - t^2}{1 + t^2}\right)^q.$$

But for any other rational number k no such rational functions exist. For consider $y = (1 - x^k)^{1/k}$ and the integral

$$\int (1 - x^k)^{1/k}\, dx = \int y dx. \tag{193}$$

If $x = f(t)$, and $y = g(t)$, by the change of variable $x = f(t)$ the integral becomes

$$\int g(t) \frac{df}{dt}\, dt$$

and since this integrand is a rational function, the integral is elementary. But, by Eqs. (189), (193) and (190), we must have

$$\frac{1}{k} \qquad \text{or} \qquad \frac{1}{k} \qquad \text{or} \qquad \frac{2}{k}$$

an integer, say q. Therefore we must have

$$k = 1/q \qquad \text{or} \qquad k = 2/q$$

and this condition is not only sufficient, but also necessary. In particular

$$k \neq 3, 4, 5, \cdots.$$

Now it is clear that if k could be an integer >2, Eq. (192) would have infinitely many rational solutions (by choosing *any* rational t) and thus

$$a^k + b^k = c^k$$

would have infinitely many integral solutions. But although Eq. (192) is not solvable in rational *functions*, this does not preclude, at least according to any known argument, a solution in terms of rational *numbers*. Although the existence of such rational functions would disprove Conjecture 16, their nonexistence does not prove it. The approach here is therefore only suggestive—it proves nothing about Conjecture 16, but it does show the special role of $n = 1$ and $n = 2$.

Three comments of general mathematical interest are these.

(a) The use of the transformations, Eq. (191), for rationalizing integrands involving $y = \sqrt{1 - x^2}$ is well known to calculus students. We see here the intimate connection with Pythagorean numbers.

(b) The reader notes that we have not previously used methods involving functions and integration, and may well ask, "What have these to do with *number* theory?" The question is well taken and in fact it may be stated that here, at least, the influence really goes in the opposite direction. The proof of Chebyshev's result, Eq. (190)—see Ritt, *Integration in Finite Terms*, Columbia University Press, 1948, p. 37—is based on certain characterizations of the algebraic functions $x^U(A + Bx^V)^W$ in terms of *integers*—namely, the number and order of the so-called branch points. It is not so much that algebraic functions have number-theoretic implications as that numbers have functional implications.

(c) We are impressed here with the fact that although Conjecture 16 has so far resisted all attempts at proof, the analogous theorem in terms of *functions* is relatively easy. There are other examples of this phenomenon in number theory. For example, there is a theorem analogous to Artin's Conjecture 13 which concerns functions, not numbers, and this has been proven by Bilharz. It would take us too far afield to elaborate.

49. The Easy Case and Infinite Descent

To prove Conjecture 16 it would clearly suffice to restrict the variables in

$$a^n + b^n = c^n \qquad (n > 2)$$

as follows:

(A) a, b, and c are prime to each other, and
(B) $n = 4$ or $n = p$, an odd prime.

For if $(a, b) = s > 1$ we may proceed as on page 141 in the proof of Theorem 59, and if $n \neq 4$ or p, it equals $4k$ or pk for some $k > 1$. But

$$a^n + b^n = c^n$$

is then impossible if

$$(a^k)^4 + (b^k)^4 = (c^k)^4$$

and

$$(a^k)^p + (b^k)^p = (c^k)^p$$

are impossible.

The only easy case is $n = 4$, and therefore also $n = 4k$. The impossibility of this case we now prove. The proof is similar to that which Fermat gave for Eq. (183), and this latter proof is noteworthy in two ways:

(A) Of all Fermat's theorems this is the *only* one for which his proof is known.

(B) The proof uses "infinite descent," a method Fermat recommended highly, which he used both for negative propositions such as Theorem 62, and with some modification, for positive propositions such as Theorem 60.

PROOF OF THEOREM 62. Assume

$$A^4 + B^4 = C^2 \tag{194}$$

where A, B, and C are prime to each other, and, without loss of generality, let A be even. Then, by Theorem 59, Eq. (181), we have

$$A^2 = 2uv, \qquad B^2 = u^2 - v^2, \qquad \text{and} \qquad C = u^2 + v^2$$

with u prime to v. Then $B^2 + v^2 = u^2$, and since B is odd, v is even. Thus

$$v = 2rs \quad \text{and} \quad u = r^2 + s^2$$

with r prime to s. Since

$$A^2 = 2uv = 4rs(r^2 + s^2),$$

by Theorem 7, r, s, and $u = r^2 + s^2$ must all be perfect squares. Let

$$r = \alpha^2, \qquad s = \beta^2, \qquad \text{and} \qquad u = \gamma^2.$$

Then

$$\alpha^4 + \beta^4 = \gamma^2 \tag{195}$$

with

$$\gamma \leqq u < C.$$

Given a solution, Eq. (194), we could thus find a second solution, Eq. (195), whose right side is smaller. But this implies an infinite sequence of positive integers

$$C > \gamma > \gamma_1 > \gamma_2 > \cdots > 0.$$

Since this infinite descent is impossible, there is no solution.

We now analyze this proof for any light it may throw on Conjecture 16, and note three features:

(A) The proof leans heavily on Theorem 59, but this is possible only because $4 = 2 \cdot 2$ and thus is not extendable to odd exponents.

(B) As in Theorems 56, 57 and 59, the unique factorization of Theorem 7 plays an important role—in distinction to, say, Chapter II, where the "Fundamental" Theorem was hardly used at all. We should expect unique factorization to be important for Conjecture 16.

(C) The infinite-descent strategy, like point B, is not peculiar to $n = 4$, and we may expect it to be useful for Conjecture 16.

Despite its rather exotic name it should be noted that infinite descent is essentially the *Well-Ordering Principle*, i.e., every nonempty set of positive integers contains a *smallest* member. As is well known, this principle is equivalent to the principle of induction—and thus is the most characteristic of all the laws concerning the integers. The reader may note that in the proof of Theorem 7 itself (page 6), and of the underlying Theorem 5 (page 9), the Well-Ordering Principle is used several times.

50. Gaussian Integers and Two Applications

To attempt Conjecture 16, the analysis above suggests that we utilize points B and C there while dropping point A. We introduce this possibility by returning first to the paradox raised on page 143. Given a prime p of the form $4m + 1$, and given, by Theorem 60,

$$p = a^2 + b^2, \tag{196}$$

we repeat, "What has the multiplicative concept of primality to do with a *sum* of two squares?" We can write Eq. (196) in a purely multiplicative manner:

$$p = (a + bi)(a - bi) \tag{197}$$

where $i = \sqrt{-1}$ and p is a *product* of the two complex factors. This is a rather ironic solution of the paradox, since in terms of these factors p is no longer a prime!

Definition 38. *Gaussian integers* are numbers of the form $a + bi$, where a and b are integers.

Examples:

$$2 + 3i, \qquad 4 - 7i, \qquad -2i, \qquad 7$$

We will give only a brief sketch of these integers. The sum, difference, and product of two Gaussian integers are Gaussian integers, but

$$a + bi|c + di$$

only if there is an $e + fi$ such that

$$(a + bi)(e + fi) = (ae - bf) + (af + be)i = c + di.$$

That is, the e and f obtained by solving $ae - bf = c$ and $af + be = d$ must both be integers. A *unity* is a divisor of 1, namely, 1, -1, i, or $-i$. Two numbers are *associated* if their ratio is a unity. A *prime* is not a unity, and is divisible only by associates of 1 and of itself. *Prime to each other* means having no common divisor other than a unity.

Consider *all* ordinary integers, positive, negative, and zero. Let $+1$ and -1 be the unities. Let a and $-a$ be associated, and let

$$\pm 2, \qquad \pm 3, \qquad \pm 5, \qquad \text{etc.}$$

be the primes. The fundamental theorem (Theorem 7) may be extended to *all* integers as follows:

Theorem 63. *Each integer not zero or a unity can be factored into a product of primes which is unique except for a possible rearrangement, and except for a possible substitution of associated primes.*

EXAMPLE:

$$-15 = 3(-5) = 5(-3), \qquad \text{etc.}$$

Now we state, without proof, that Theorem 63 is also true for Gaussian integers. Assuming this, we will give two applications.

PARTIAL PROOF OF THEOREM 60. On the basis of this unique factorization we will show that every $p = 4m + 1$ is a sum of two squares. Since $(-1 \mid p) = 1$ there is an s such that

$$p|s^2 + 1.$$

Let the quotient be q and

$$pq = s^2 + 1 = (s + i)(s - i).$$

Now p cannot be a Gaussian prime, for if it were, by the unique factorization of $s^2 + 1$, we would have $p|s + i$ or $p|s - i$. Since neither quotient is a Gaussian integer, p is not a Gaussian prime. But it is not divisible by a real prime. Therefore

$$a + bi|p$$

for some a and b. Since p is real the quotient must be $c(a - bi)$. Thus $p = c(a^2 + b^2)$, and since c must be 1, we have $p = a^2 + b^2$.

PARTIAL PROOF THAT EQ. (183) IS IMPOSSIBLE. Assume

$$A^4 - B^4 = C^2, \qquad B \equiv 0 \pmod 2 \tag{198}$$

with A, B, and C prime to each other. Then C is odd and

$$A^4 = (C + iB^2)(C - iB^2).$$

By unique factorization $C + iB^2$ is *associated* with a perfect fourth power. Assume first that it equals a fourth power or its negative:

$$C + iB^2 = (D + iE)^4 \quad \text{or} \quad = -(E + iD)^4.$$

Then

$$B^2 = 4DE(D^2 - E^2), \qquad \pm C = D^4 - 6D^2E^2 + E^4.$$

Since D, E, and $D^2 - E^2$ are prime to each other they are perfect squares. Let $D = \alpha^2$, $E = \beta^2$, and $D^2 - E^2 = \gamma^2$. Then

$$\alpha^4 - \beta^4 = \gamma^2, \tag{198a}$$

and, since $4\beta^2 | B^2$, $\beta < B$. Since C is odd, D and E are not both odd. And, since $D^2 - E^2$ is a square $= 4m + 1$, we must have E and therefore β even. Finally, $C + iB^2 \neq \pm i(D + iE)^4$, since equality here implies that C is even. Then, by infinite descent, from Eq. (198a) and $\beta \equiv 0 \pmod 2$, Eq. (198) is impossible.

By a somewhat similar approach, using a generalized unique factorization, and infinite descent, we now examine Conjecture 16.

EXERCISE 108. Show that Bachet's problem (page 144) is equivalent to Eq. (198) and therefore impossible.

51. ALGEBRAIC INTEGERS AND KUMMER'S THEOREM

We generalize Gaussian integers and sketch the following. A root z of a polynomial with integer coefficients is called an algebraic number. The set of all numbers

$$w = \frac{f(z)}{g(z)}$$

which are rational functions (with integer coefficients) of z is called an *algebraic number field*, $k(z)$. The numbers of such a field which are roots of a polynomial

$$w^n + aw^{n-1} + \cdots + s = 0$$

with integer coefficients, and leading coefficient 1, are the *algebraic integers* of that field.

EXAMPLE:

The Gaussian integers are the algebraic integers of $k(\sqrt{-1})$ for if a and b are ordinary integers, $a + bi$ is a root of

$$w^2 - 2aw + a^2 + b^2 = 0,$$

and it may be shown that all other numbers in $k(\sqrt{-1})$ are not roots of a polynomial with leading coefficient 1.

Unities, associated numbers, and primes are defined as before. If Theorem 63 held for the algebraic integers of any field then Conjecture 16 could be shown to follow.

Assume

$$A^p + B^p = C^p \tag{199}$$

with A, B, and C prime to each other, and p an odd prime. Let

$$\rho = e^{2\pi i/p} \tag{200}$$

and we may then factor the left side of Eq. (199) as follows

$$(A + B)(A + \rho B)(A + \rho^2 B) \cdots (A + \rho^{p-1}B) = C^p. \tag{201}$$

It may be shown that the algebraic integers of $k(\rho)$ are

$$a + b\rho + c\rho^2 + \cdots + s\rho^{p-2}$$

where a, b, $\cdots$ are ordinary integers.

We have, therefore, as in Eq. (197), turned an additive problem into a purely multiplicative problem.

Now *if* these algebraic integers had unique factorization we could deduce from Eq. (201) that each factor on the left is associated with a perfect pth power of an algebraic integer. If this were always true, Fermat's Last Theorem would follow. E. E. Kummer, A. L. Cauchy and G. Lamé all assumed that such uniqueness did exist. However, Dirichlet pointed out that this must be proven. In fact, it is *not* true in general—the first counter-example being $p = 23$. To overcome this lack of unique factorization into primes, Kummer was led to introduce the important, underlying *ideal numbers*, a development we cannot enter into here.

With this theory Kummer obtained a proof of Conjecture 16 for many prime exponents n. We will state his remarkable result but not attempt to prove it, as the proof is long and difficult.

Definition 39. The *Bernoulli number* B_n is a rational number defined by the power series:

$$\frac{x}{e^x - 1} = 1 - \frac{x}{2} + \sum_{n=1}^{\infty} (-1)^{n-1} \frac{B_n x^{2n}}{(2n)!}. \tag{202}$$

EXAMPLES:

$$B_1 = \frac{1}{6}, \qquad B_2 = \frac{1}{30}, \qquad B_3 = \frac{1}{42}, \cdots, \qquad B_{16} = \frac{7709321041217}{510}$$

Definition 40. A prime p is *regular* if it divides none of the numerators of

$$B_1, B_2, B_3, \cdots, B_{(p-3)/2}$$

when these numbers are written in their lowest terms. Otherwise p is *irregular.*

EXAMPLE:

Since $37 | 7709321041217$, and the larger number is the numerator of B_{16}, and $16 \leqq \frac{1}{2}(37 - 3)$, 37 is irregular.

Theorem 64 (Kummer, 1850). *Fermat's Last Theorem is true for every exponent which is a regular prime. The only irregular primes up to* 100 *are* 37, 59, *and* 67.

COMMENTS:

(A) The definition of regular is explicit but complicated; it has no *apparent* relation to the problem. There is a more basic definition in terms of the so-called *class number* but this is less explicit numerically and would take longer to explain. This latter concept is fundamental, but is beyond our scope.

(B) The name "irregular" is really misleading. Although only 3 of the first 24 odd primes, $2 < p < 100$, are irregular, larger primes are "irregular" more often. Of the 367 primes, $2 < p < 2520$, 144 are irregular; and of the next 183 primes, $2520 < p < 4002$, 72 are irregular. These ratios:

$$\frac{144}{367} = .392 \quad \text{and} \quad \frac{72}{183} = .393,$$

are substantial.

(C) Other criteria have been found, besides Theorem 64, and applied to the irregular prime cases. With the aid of the SWAC, Selfridge, Nicol and Vandiver proved that Conjecture 16 is true for all exponents $\leqq 4002$. But with Kummer's regular primes, and other primes allowed by other known criteria, it has not yet been *proven* that there are infinitely many valid prime exponents.

Before leaving nonunique factorization let us examine a few examples. Consider the *quadratic* fields $k(\sqrt{N})$ where, without loss of generality, N is quadratfrei. Of the 12 cases, $N = -7, -6, -5, -3, -2, -1, 1, 2, 3, 5, 6, 7$, only in $k(\sqrt{-5})$ and $k(\sqrt{-6})$ do the integers not have unique factorization. We show two well-known examples. In $k(\sqrt{-5})$,

$$21 = 3 \cdot 7 = (1 + 2\sqrt{-5})(1 - 2\sqrt{-5})$$

although the four factors here may all be shown to be primes. In $k(\sqrt{-6})$,

$$6 = -\sqrt{-6}(\sqrt{-6}) = 2 \cdot 3$$

and again the factors are primes. Finally we note, in passing, that in the corresponding two quadratic forms,

$$a^2 + 5b^2 = (a + \sqrt{-5}b)(a - \sqrt{-5}b),$$

$$a^2 + 6b^2 = (a + \sqrt{-6}b)(a - \sqrt{-6}b),$$

if we set $a = n$ and $b = 1$, and consider

$$n^2 + 5 \quad \text{and} \quad n^2 + 6$$

we obtain forms with an exceptionally low density of primes. See the table on page 49. This is not a coincidence—the low density is really related to the nonunique factorization—but the argument is a long one.

52. The Restricted Case, Sophie Germain, and Wieferich

Sophie Germain, a Parisian lady, was a contemporary of Gauss. Since the École Polytechnique did not accept women in the school she took correspondence courses. She wrote Gauss after his *Disquisitiones* appeared telling him how much she liked the book. She included some of her own discoveries and signed—as talented ladies did in those days—with a male pseudonym, "M. Le Blanc." Gauss was impressed. Only later, under interesting circumstances, did he learn that M. Le Blanc was a lady. Gauss was astonished and pleased. Henceforth their correspondence was not strictly technical; he told her his birthday, etc.

There is a special case of Conjecture 16 which is substantially easier. In this, the *restricted* case, it is assumed, in

$$A^p + B^p = C^p$$

that p does not divide A, B, or C. Since it is possible that this case is true while Conjecture 16 is false we state it separately.

Conjecture 17. *The equation*

$$a^p + b^p = c^p$$

has no solution in integers not divisible by p.

The *Encyclopaedia Britannica* (1960) states erroneously that Sophie Germain proved this conjecture; the article should add: for $p < 100$. We give a cut-down version of her result. It shows

(A) How far one can go with very elementary arguments,
(B) That the restricted case is indeed easier, and
(C) There is a relation to our Conjecture 5. See pages 30 and 31.

Theorem 65 (Germain, modified). *The equation*

$$A^p + B^p = C^p \tag{203}$$

has no solution in integers prime to p if p is an odd prime, and $q = 2p + 1$ *is also a prime.*

Proof. Assume a solution. We may take A, B, and C prime to each other, and since p is odd we may write Eq. (203) symmetrically:

$$R^p + S^p + T^p = 0 \tag{204}$$

where $A = R$, $B = S$, and $-C = T$. Consider

$$S^p + T^p = (-R)^p.$$

Both sides are divisible by $S + T$ by Theorem 4_0 on page 17, and since $p \nmid R$, we obtain $p \nmid S + T$. Now let $m = 1$ in Theorem 48, page 105, and let

$$S = x = a \quad \text{and} \quad T = -y = -b.$$

Since $p \nmid S + T = a - b$, by Eq. (141), $S + T = x - y$ is prime to

$$z = \frac{S^p + T^p}{S + T} = \frac{(-R)^p}{S + T}.$$

Therefore, since

$$(-R)^p = \frac{(-R)^p}{S + T} \cdot (S + T),$$

both factors on the right are perfect pth powers. Write

$$S + T = r^p, \qquad \frac{S^p + T^p}{S + T} = \rho^p, \qquad -R = r\rho. \tag{205}$$

Similarly, by symmetry,

$$\begin{aligned} T + R &= s^p, \qquad \frac{T^p + R^p}{T + R} = \sigma^p, \qquad -S = s\sigma. \\ R + S &= t^p, \qquad \frac{R^p + S^p}{R + S} = \tau^p, \qquad -T = t\tau. \end{aligned} \tag{206}$$

Therefore

$$2R = s^p + t^p - r^p. \tag{207}$$

Now, by Euler's Criterion, if $q = 2p + 1$ does not divide R, S, or T, we have

$$R^p, S^p, T^p \quad \text{all} \quad \equiv \pm 1 \pmod{q}.$$

By Eq. (204) this is impossible. On the other hand, q cannot divide two or three of R, S, and T since they are prime to each other. Therefore q divides exactly one of them. Let it be R. From Eq. (207) it therefore follows similarly that q divides exactly one of r, s, and t, and by Eq. (206) it must be r.

Then, since $q \mid R$, from

$$\sigma^p(T + R) = T^p + R^p$$

we have

$$\sigma^p T \equiv T^p \pmod{q},$$

or

$$\sigma^p \equiv T^{p-1} \pmod{q}.$$

Now, since $q \nmid \sigma$,

$$T^{p-1} \equiv \pm 1 \pmod{q}.$$

From Eq. (205), $S \equiv -T \pmod{q}$, and therefore, from

$$\rho^p = \frac{S^p + T^p}{S + T} = S^{p-1} - S^{p-2}T + \cdots + T^{p-1},$$

$$\rho^p \equiv pS^{p-1} \equiv pT^{p-1} \equiv \pm p \pmod{q}.$$

This is impossible, since $(\rho \mid q) = \pm 1$, and by Euler's Criterion the theorem is proven.

COMMENTS:

(A) Therefore Conjecture 17 is true for $p = 3, 5, 11, 23, \cdots, 16188302111, \cdots$.

(B) If Conjecture 5 were true, Conjecture 17 would be true for infinitely many primes. But the latter has never been proven.

(C) By a modification of the argument, the criterion:

$$2p + 1 \text{ is a prime,}$$

may be supplemented by other criteria. It suffices if any of the following are true:

$$4p + 1, \quad 8p + 1, \quad 10p + 1, \quad 14p + 1, \quad \text{or} \quad 16p + 1$$

is a prime. For example, since

$$29 = 4 \cdot 7 + 1 \text{ is a prime,}$$

Conjecture 17 is true for $p = 7$. The above criteria, taken together, suffice for all $p < 100$. Therefore, as S. Germain proved, Conjecture 17 is true for all $p < 100$.

(D) Theorem 65 has about the easiest proof of any significant result obtained on Fermat's Last Theorem. That the restricted case is much easier is also shown by the fact that in Kummer's Theorem 64 the greatest difficulty comes when the restriction is waived. Further, as we shall soon see, there can be little doubt that Conjecture 17 is true. Still, it has not been proven—not even for infinitely many p—as already stated in point (B).

(E) Unique factorization is again fundamental. (Where does it enter?)

In 1909 A. Wieferich showed that Conjecture 17 is true if

$$p^2 \nmid 2^{p-1} - 1, \tag{208}$$

that is, if p^2 is *not* a "Wieferich square" (see pages 116, 118). This criterion is therefore sufficient for all $p < 100{,}000$ except 1093 and 3511. However, despite the fact that these squares are so rare, no one has proven that there are infinitely many p which satisfy (208). Further, D. Mirimanoff subsequently (1910) showed that

$$p^2 \nmid 3^{p-1} - 1 \tag{209}$$

was an equally valid criterion. Therefore the rare Wieferich squares must also violate the equally prevalent (209) if we are to discover a counter-example for Conjecture 17.

With these and other similar criteria, and following many previous authors, D. H. and Emma Lehmer showed that Conjecture 17 is true for all primes $<253{,}747{,}889$.

Exercise 109. Show that Conjecture 17 is true for $p = 3$ since if $3 \nmid a$, $a^3 \equiv \pm 1 \pmod 9$.

Exercise 110. Show that the 24 odd primes <100 satisfy one or another of the six criteria in point (C), page 156.

Exercise 111. If Conjecture 17 is true for all prime exponents does it follow that it is true for all exponents, as on page 148?

53. Euler's "Conjecture"

Although

$$A^3 + B^3 \neq C^3$$

in positive integers we do have

$$6^3 = 3^3 + 4^3 + 5^3 \quad \text{and} \tag{210}$$

$$29^3 = 11^3 + 15^3 + 27^3. \tag{211}$$

There are in fact infinitely many solutions of

$$D^3 = A^3 + B^3 + C^3. \tag{212}$$

In our proof of

$$A^4 + B^4 \neq C^4$$

we utilized the fact that we had *all* solutions of

$$a^2 + b^2 = c^2$$

and that, of these, a and b could not be squares simultaneously. The strategy suggests itself to find all solutions of Eq. (212) and, by specialization, to show that Fermat's Last Theorem is true for $n = 3$. Further, one could hope for a similar approach to $n = 5, 7$, etc.

We know of no serious progress in this direction. In this connection there is a "conjecture" of Euler. While it has an attractive ring to it we know of no serious evidence and so shall call it

Open Question 2. *Can an* nth *power ever equal the sum of fewer than* n nth *powers? That is, can*

$$A^n = B_1^n + B_2^n + \cdots + B_k^n$$

for $1 < k < n$?

Euler "conjectured" no. If his "conjecture" were true, Fermat's Last Theorem would follow as a special case.

EXERCISE 112 (From Dickson). Write Eq. (212) in symmetrized form

$$W^3 + X^3 + Y^3 + Z^3 = 0. \tag{213}$$

Substitute

$$W = \tfrac{1}{2}(w + x + y + z), \qquad X = \tfrac{1}{2}(w + x - y - z)$$
$$Y = \tfrac{1}{2}(w - x + y - z), \qquad Z = \tfrac{1}{2}(w - x - y + z)$$

and show that Eq. (213) becomes the determinantal equation:

$$\begin{vmatrix} w & 3z & -3y \\ -z & w & 3x \\ y & -x & w \end{vmatrix} = 0. \tag{214}$$

This is the condition that

$$wa + 3zb - 3yc = 0, \qquad -za + wb + 3xc = 0, \qquad ya - xb + wc = 0$$

have solutions a, b, and c not all zero.

Solve for x, y, and z in terms of w and obtain solutions of Eq. (214):

$$w = -6\rho abc \qquad\qquad x = \rho a(a^2 + 3b^2 + 3c^2)$$
$$y = \rho b(a^2 + 3b^2 + 9c^2) \qquad z = 3\rho c(a^2 + b^2 + 3c^2).$$

Now with $a = b = c = 1$ and a proper choice of ρ obtain Eq. (211). Conversely, from Eq. (210), obtain an a, b, c, and ρ which gives that solution. Finally, can all solutions of Eq. (212) in integers be obtained by these formulae?

54. Sum of Two Squares

On pages 150–151 we gave a nonconstructive, partial proof of Theorem 60 based upon Gaussian integers. We now give two complete proofs, the first explicitly constructive, and the second implicitly constructive. Both are based upon $(-1|p) = +1$. The first proof—it may be Fermat's—uses the method of descent and also a famous identity which goes back (at least) to Diophantus:

Theorem 66.

$$\begin{aligned}(a^2 + b^2)(c^2 + d^2) &= (ac + bd)^2 + (ad - bc)^2 \\ &= (ac - bd)^2 + (ad + bc)^2.\end{aligned} \tag{215}$$

Proof. Clear.

Proof of Theorem 60. If p is a prime $\equiv 1 \pmod 4$, there is an $s < p$ such that $p|s^2 + 1$. Write $s = a_0$, $1 = b_0$, and

$$pq_0 = a_0^2 + b_0^2. \tag{216}$$

It follows that $q_0 < p$. If $q_0 = 1$, $p = a_0^2 + b_0^2$. If not, divide a_0 and b_0 by q_0 choosing remainders, positive or negative, which have a minimum magnitude:

$$a_0 = r_0q_0 + \alpha_0, \qquad b_0 = s_0q_0 + \beta_0 \tag{217}$$

Both remainders, α_0 and β_0, therefore satisfy $|x| \leqq \frac{1}{2}q_0$, and not both are zero. For if $\alpha_0 = \beta_0 = 0$, we have $q_0|p$, and since $1 < q_0 < p$ this is impossible.

Now define q_1 by

$$q_0q_1 = \alpha_0^2 + \beta_0^2 \tag{218}$$

and we have $0 < q_1 \leqq \frac{1}{2}q_0$. But, from Eqs. (216) and (215), we have

$$\begin{aligned}pq_0^2q_1 &= (a_0^2 + b_0^2)(\alpha_0^2 + \beta_0^2) \\ &= (a_0\alpha_0 + b_0\beta_0)^2 + (a_0\beta_0 - b_0\alpha_0)^2.\end{aligned}$$

Substituting Eq. (217), and dividing by q_0^2, now yields

$$pq_1 = (r_0\alpha_0 + s_0\beta_0 + q_1)^2 + (r_0\beta_0 - s_0\alpha_0)^2. \tag{219}$$

Thus, if

$$\begin{aligned}a_1 &= |r_0\alpha_0 + s_0\beta_0 + q_1| \\ b_1 &= |r_0\beta_0 - s_0\alpha_0|\end{aligned} \tag{220}$$

we have

$$pq_1 = a_1^2 + b_1^2. \tag{221}$$

If $q_1 \neq 1$, we continue, and obtain

$$q_0 > q_1 > \cdots > q_n = 1.$$

Finally

$$p = a_n^2 + b_n^2.$$

To show the uniqueness asserted in Theorem 60, we assume a, b, c, and d are positive and

$$p = a^2 + b^2 = c^2 + d^2. \tag{222}$$

By Eq. (215),

$$p^2 = (ac + bd)^2 + (ad - bc)^2 \tag{223}$$

and

$$p^2 = (ac - bd)^2 + (ad + bc)^2. \tag{224}$$

By Eq. (222), $(p - a^2)\, d^2 = (p - c^2)b^2$ or

$$p(d^2 - b^2) = (ad - bc)(ad + bc).$$

Now if $p|ad - bc$, by Eq. (223) we have $ad - bc = 0$, and thus $d^2 - b^2 = 0$ or $d = b$. Whereas, if $p|ad + bc$, by Eq. (224) we have $ac = bd$, and since $(a, b) = 1$, we have $a|d$ and $b|c$. By Eq. (222) we now have $d = a$. Since p is prime, we must have one of these two cases.

Finally, to make the determination of $p = a_n^2 + b_n^2$ completely constructive—but not necessarily efficient—we note that we may take

$$s = a_0 \equiv \left(\frac{p-1}{2}\right)! \qquad (\text{mod } p)$$

by Wilson's Theorem, and Exercise 22, page 38.

Exercise 113. Determine $29 = a_1^2 + b_1^2$ and $89 = a_2^2 + b_2^2$, given $29 \mid 12^2 + 1$ and $89 \mid 34^2 + 1$.

Exercise 114. From the previous exercise find the two representations of

$$29 \cdot 89 = 2581 = A^2 + B^2$$

using Eq. (215).

Exercise 115. Given $p = a^2 + b^2$, determine $2p = A^2 + B^2$, and $5p = C^2 + D^2 = E^2 + F^2$.

Exercise 116. Using the results of Exercise 113, find, conversely, an x and y such that $29|x^2 + 1$, and $89|y^2 + 1$ by $x \equiv a_1b_1^{-1}$ (mod 29), and $y \equiv a_2b_2^{-1}$ (mod 89).

A shorter, more modern proof of Theorem 60 is related to the idea in Exercise 116. It uses *Thue's Theorem*, and this, in turn, uses the

Dirichlet Box Principle. *If more than N objects are placed in N boxes, at least one box contains two or more objects.*

Theorem 67 (Thue). *If* $n > 1$, $(a, n) = 1$, *and* m *is the least integer* $> \sqrt{n}$, *there exist an* x *and* y *such that*

$$ay \equiv +x \quad \text{or} \quad -x \qquad (\text{mod } n)$$

where

$$0 < x < m, \qquad 0 < y < m.$$

Proof. Consider $ay - x$ for the m^2 possibilities: $y = 0, 1, 2, \cdots, m - 1$ and $x = 0, 1, 2, \cdots, m - 1$. Since $m^2 > n$, by the Dirichlet Box Principle at least two of these possibilities must be congruent modulo n. Let

$$ay_1 - x_1 \equiv ay_2 - x_2 \qquad (\text{mod } n)$$

with $y_1 > y_2$. Further $x_1 \neq x_2$, for otherwise, since $(a, n) = 1$, we have $y_1 = y_2$. Let $y = y_1 - y_2$ and $x = \pm(x_1 - x_2) > 0$ and we have

$$ay \equiv +x \quad \text{or} \quad -x \qquad (\text{mod } n)$$

as required.

Second Proof of Theorem 60. Let $p|s^2 + 1$. By Thue's Theorem there exist positive integers x and $y < \sqrt{p}$ such that

$$ys \equiv \pm x \qquad (\text{mod } p).$$

Since $(y, p) = 1$, we have

$$s^2 + 1 \equiv x^2y^{-2} + 1 \equiv 0 \qquad (\text{mod } p)$$

or

$$x^2 + y^2 \equiv 0 \qquad (\text{mod } p).$$

But $0 < x^2 + y^2 < 2p$. Therefore $p = x^2 + y^2$. The uniqueness we prove as before.

Exercise 117. Apply the Dirichlet Box Principle to Gertrude Stein's surrealist opera, *Four Saints in Three Acts*, and draw a valid inference.

55. A Generalization and Geometric Number Theory

Fermat, in a letter to Frenicle (1641), called Theorem 60 "the fundamental theorem on right triangles." Compounding factors by Eq. (215), he obtained numerous results such as:

A prime $= 4m + 1$ is the hypothenuse of a Pythagorean triangle in a single way, its square in two ways, its cube in three ways, etc.

EXAMPLE:

$$5^2 = 4^2 + 3^2$$

$$25^2 = 24^2 + 7^2 = 20^2 + 15^2$$

$$125^2 = 120^2 + 35^2 = 117^2 + 44^2 = 100^2 + 75^2, \quad \text{etc.}$$

It is clear, from Eq. (215), that the product of two distinct primes of the form $4m + 1$ is a hypothenuse in two ways, and, it may be shown, that a product of k such primes is a hypothenuse in 2^{k-1} ways.

EXERCISE 118. Obtain 4 distinct representations of $n = A^2 + B^2$ for (the Carmichael number) $n = 5 \cdot 13 \cdot 17 = 1105$.

We asked, on page 143: In how many ways is n a sum of two squares? The answer takes a particularly neat form if we alter the convention of what we mean by "how many ways."

Definition 41. By $r(n)$ we mean the number of representations $n = x^2 + y^2$ in integers x and y, which are positive, negative, or zero. The representations are considered distinct even if the x's and y's differ only in sign or order. Further we define $R(N)$ by

$$R(N) = \sum_{n=0}^{N} r(n). \tag{225}$$

EXAMPLES:

$$r(0) = 1 \quad \text{since} \quad 0 = 0^2 + 0^2.$$

$$r(4) = 4 \quad \text{since} \quad 4 = (\pm 2)^2 + 0^2 = 0^2 + (\pm 2)^2.$$

$$r(8) = 4 \quad \text{since} \quad 8 = (\pm 2)^2 + (\pm 2)^2.$$

$$r(10) = 8 \quad \text{since} \quad 10 = (\pm 1)^2 + (\pm 3)^2 = (\pm 3)^2 + (\pm 1)^2.$$

$$r(p) = 8 \text{ if } p \text{ is a prime } = 4m + 1.$$

$$R(12) = 1 + 4 + 4 + 0 + \cdots + 0 = 37.$$

It can be shown, by elementary methods, that the following result holds.

Theorem 68. *If n, $\geqq 1$, has A positive divisors $\equiv 1 (mod\ 4)$ and B positive divisors $\equiv -1 (mod\ 4)$, then*

$$r(n) = 4(A - B). \tag{226}$$

We mean here all divisors, not merely prime divisors.

EXAMPLES:

$r(2) = 4$ since $A = 1$; (1). $B = 0$.

$r(5) = 8$ since $A = 2$; (1, 5). $B = 0$.

$r(7) = 0$ since $A = 1$; (1). $B = 1$; (7).

$r(65) = 16$ since $A = 4$; (1, 5, 13, 65). $B = 0$.

Theorem 68 contains Theorem 60 as a special case when allowance is made for the different conventions. We now apply this generalization to derive the famous Leibnitz series:

$$\tfrac{1}{4}\pi = 1 - \tfrac{1}{3} + \tfrac{1}{5} - \tfrac{1}{7} + \tfrac{1}{9} - \cdots. \tag{227}$$

Equation (227) was one of the first results obtained by Leibnitz from his newly discovered integral calculus. In the subsequent priority controversy concerning the calculus, Newton's supporters pointed out that Gregory had already given

$$\arctan x = x - \tfrac{1}{3}x^3 + \tfrac{1}{5}x^5 - \cdots$$

and Eq. (227) follows by taking $x = 1$. Our present interest concerns quite a different point—a remark by Leibnitz concerning Eq. (227). He suggested that with Eq. (227) he had reduced the mysterious number π to the integers. We may contest this claim. The derivation of Eq. (227) using integration and Taylor's series does not reveal the number-theoretic relation between π and the odd numbers. One may ask, "What has a circle to do with odd numbers?" and receive no convincing answer from this derivation. The real insight is given by Theorem 68.

Consider the number of Cartesian lattice points (a, b) in or on the circle $x^2 + y^2 = N$. We show these points for $N = 12$. There are 37 of them.

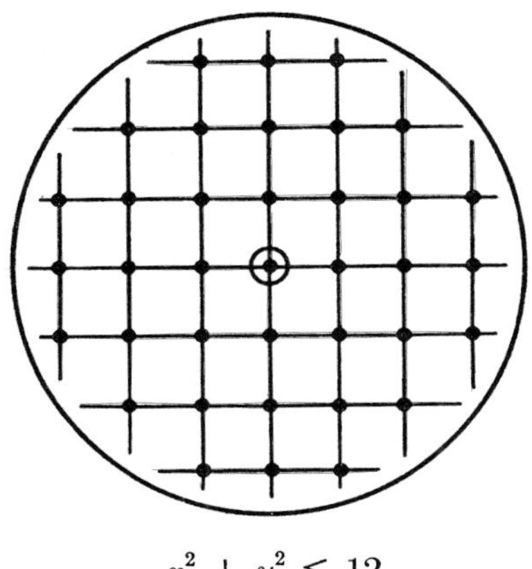

$x^2 + y^2 \leqq 12$

It is clear, by Definition 41, that the number of such points equals $R(N)$, since each point corresponds to exactly one representation of one $n = a^2 + b^2 \leqq N$. Further, if we associate each point (a, b) with the unit square of which it is the center, $(a \pm \frac{1}{2}, b \pm \frac{1}{2})$, we see that $R(N)$ approximates the area of the circle, πN. The reader may show that the difference, $R(N) - \pi N$, vanishes with respect to πN as $N \to \infty$, since this difference is associated with the (relatively small) region along the circumference. In this way he will obtain

Theorem 69.

$$R(N) \sim \pi N. \tag{228}$$

Corollary. *The mean number of representations of $n = a^2 + b^2$, for n up to $n = N$, tends to π as $N \longrightarrow \infty$.*

But, from Theorem 68, we may obtain a neat and exact formula for $R(N)$. Each $n \leqq N$ receives a contribution of 4 representations from its divisor 1. Each $n \leqq N$, which is divisible by 3, loses 4 representations from this divisor 3, and there are $[N/3]$ such values of n. Similarly, there are $4[N/(2k + 1)]$ contributions, or $4[N/(2k + 1)]$ losses, corresponding to the odd divisor $2k + 1$, according as k is even or odd. Counting the single representation of $0 = 0^2 + 0^2$, we thus obtain

$$R(N) = 1 + 4\left\{\left[\frac{N}{1}\right] - \left[\frac{N}{3}\right] + \left[\frac{N}{5}\right] - \left[\frac{N}{7}\right] + \cdots\right\}. \tag{229}$$

Further, since $[N/(2k + 1)] = 0$ if $2k + 1 > N$, we may write Eq. (229) as an infinite series, and thus obtain

Theorem 70.

$$R(N) = 1 + 4\sum_{k=0}^{\infty} (-1)^k \left[\frac{N}{2k + 1}\right]. \tag{230}$$

Example:

$$R(12) = 1 + 4\{12 - 4 + 2 - 1 + 1 - 1\} = 37.$$

With Theorem 69, dropping the 1 as $N \to \infty$, we obtain

$$\tfrac{1}{4}\pi N \sim \sum_{k=0}^{\infty} (-1)^k \left[\frac{N}{2k + 1}\right]. \tag{231}$$

Now split the right side into two sums:

$$\sum_{k=0}^{K} (-1)^k \left[\frac{N}{2k + 1}\right] + \sum_{K+1}^{\infty} (-1)^k \left[\frac{N}{2k + 1}\right]$$

where

$$K = [\sqrt{N}] - 1.$$

There are $[\sqrt{N}]$ terms in the first sum and we have

$$\sum_{k=0}^{K} (-1)^k \left[\frac{N}{2k+1}\right] = N \sum_{k=0}^{K} (-1)^k \frac{1}{2k+1} + \theta\sqrt{N},$$

where $|\theta| < 1$, since the error made by removing the square brackets in each term is <1. On the other hand the magnitude of the second sum is less than, or equal to, the magnitude of its leading term—since the terms are alternating in sign and monotonic in magnitude. Therefore it equals $\theta'\sqrt{N}$ where $|\theta'| < 1$. Therefore, dividing by N, Eq. (231) now becomes

$$\tfrac{1}{4}\pi \sim \sum_{k=0}^{K} (-1)^k \frac{1}{2k+1} + \frac{\theta''}{\sqrt{N}}$$

where $|\theta''| < 2$, and, letting $N \to \infty$, Eq. (227) follows.

Exercise 119. Gauss gave $R(100) = 317$ and $R(10{,}000) = 31417$. Verify the former, using Eq. (230).

Exercise 120. Jacobi's proof of Theorem 68 was not elementary but was based upon an identity which he obtained from elliptic functions:

$$(1 + 2x + 2x^4 + 2x^9 + 2x^{16} + \cdots)^2$$
$$= 1 + 4\left(\frac{x}{1-x} - \frac{x^3}{1-x^3} + \frac{x^5}{1-x^5} - \cdots\right).$$

Show that if the left side is written as a power series,

$$1 + a_1x + a_2x^2 + a_3x^3 + \cdots,$$

then $a_n = r(n)$, while if the right side is

$$1 + b_1x + b_2x^2 + b_3x^3 + \cdots,$$

then $b_n = 4(A - B)$, where A and B are as in Theorem 68.

56. A Generalization and Binary Quadratic Forms

We now (start to) generalize Theorem 60 in a different direction. We consider numbers of the form $x^2 + Ny^2$ as suggested on page 143. At first things go easily. Theorem 66 becomes

Theorem 71.

$$\begin{aligned}(a^2 + Nb^2)(c^2 + Nd^2) &= (ac + Nbd)^2 + N(ad - bc)^2 \\ &= (ac - Nbd)^2 + N(ad + bc)^2.\end{aligned} \tag{232}$$

PROOF. Consider

$$(a + \sqrt{-N}b)(a - \sqrt{-N}b)(c + \sqrt{-N}d)(c - \sqrt{-N}d).$$

By pairing the 1st and 2nd terms, and the 3rd and 4th we obtain the left side of Eq. (232). By pairing the 1st and 4th terms, and the 2nd and 3rd we obtain the first right side; while pairing the 1st and 3rd terms, etc., gives the second right side.

For $N = 2$ and 3, Theorem 60 generalizes easily to

Theorem 72. *Every prime p of the forms $8m + 1$ and $8m + 3$ can be written as $p = x^2 + 2y^2$ in a unique way. Every prime p of the form $6m + 1$ can be written as $p = x^2 + 3y^2$ in a unique way.*

PROOF. If $-N$ is a quadratic residue of p there is an s, prime to p, such that $s^2 + N \equiv 0 \pmod{p}$. By Thue's Theorem, as on page 161, there are positive integers x and $y < \sqrt{p}$ such that

$$s^2 + N \equiv x^2y^{-2} + N \equiv x^2 + Ny^2 \equiv 0 \pmod{p}.$$

Now if $p = 8m + 1$ or $8m + 3$, $(-2|p) = +1$, and $x^2 + 2y^2$ is a multiple of p which is $<3p$. If $x^2 + 2y^2 = p$, we have our solution; but if $x^2 + 2y^2 = 2p$, since x must be even, $= 2w$, we have $y^2 + 2w^2 = p$ as our solution.

Again, if $p = 6m + 1$, $(-3|p) = +1$, and $x^2 + 3y^2$ is a multiple of $p < 4p$. Now $x^2 + 3y^2 \neq 2p$, for if equality holds, x and y are either both odd or both even and therefore $x^2 + 3y^2$ is divisible by 4, that is, $2|p$. Therefore either $p = x^2 + 3y^2$, or $x = 3w$ and $y^2 + 3w^2 = p$, as before.

The uniqueness follows from the more general

Theorem 73. *If $N > 0$ there is at most one representation of a prime p as $p = a^2 + Nb^2$ in positive integers a and b.*

PROOF. This is left for the reader, who will utilize Theorem 71.

Now, the "natural" generalization of Theorem 72 would be this—if $(-N|p) = +1$, then $p = x^2 + Ny^2$ in a unique way—but this supposition is not true. The generalization breaks down at two points.

First, as hinted by the qualification, $N > 0$, in Theorem 73, uniqueness need not hold if $N < 0$. Thus we have (see page 143) the Fermat-Frenicle

Theorem 74. *Every prime p of the form $8m \pm 1$ can be written as $a^2 - 2b^2$ in infinitely many ways.*

PROOF. Since $(2|p) = +1$, we have, by Thue's Theorem, $x^2 - 2y^2$ is a multiple of p, $<p$ and $>-2p$. Since $x^2 - 2y^2 \neq 0$ by Theorem 56, we have

$$x^2 - 2y^2 = -p.$$

Therefore

$$(x + 2y)^2 - 2(x + y)^2 = p$$

or

$$a^2 - 2b^2 = p. \tag{233}$$

Now let a_{2n} and c_{2n} be the side and diagonal numbers of Theorem 58, page 139. Then by Eqs. (232) and (172),

$$(c_{2n}^2 - 2a_{2n}^2)(a^2 - 2b^2) = p$$

and

$$p = (c_{2n}a - 2a_{2n}b)^2 - 2(c_{2n}b - a_{2n}a)^2.$$

Likewise

$$p = (c_{2n}a + 2a_{2n}b)^2 - 2(c_{2n}b + a_{2n}a)^2.$$

Therefore from each of the infinitely many pairs (c_{2n}, a_{2n}), and from Eq. (233) we obtain two other solutions of Eq. (233).

EXAMPLE:
From $3^2 - 2 \cdot 1^2 = 7$, and $(c_{2n}, a_{2n}) = (3, 2)$ and $(17, 12)$, we find:

$$7 = 5^2 - 2 \cdot 3^2 = 13^2 - 2 \cdot 9^2 = 27^2 - 2 \cdot 19^2 = 75^2 - 2 \cdot 53^2.$$

EXERCISE 121. From $5^2 - 2 \cdot 2^2 = 17$, find four other representations of $a^2 - 2b^2 = 17$.

To generalize Theorem 74 to $a^2 - 3b^2$, $a^2 - 5b^2$, etc., we would need the generalization of Theon's Theorem 58 known as Fermat's Equation, i.e., Eq. (174). This we will investigate in Sect. 58 below. We may also note that the infinite number of solutions in Theorem 74, in distinction to the single solution in Theorem 72, is associated with the fact that the algebraic number field $k(\sqrt{2})$ has *infinitely* many unities—see pages 152, 150. That is,

$$c_n + \sqrt{2}a_n | 1$$

for any side and diagonal numbers a_n and c_n.

A second, and more difficult, point which precludes the simple generalization of Theorem 72 mentioned on page 166 is this. In the proof of Theorem 72—say with $N > 0$—one finds an x and y such that $x^2 + Ny^2 = rp$, where the coefficient r satisfies $1 \leqq r < N + 1$. It is not clear that, with these many possibilities for r, one can always obtain an $r = 1$.

Indeed, for $N = 5$ and 6 this is impossible. Thus $(-6|p) = +1$ for $p = 24m + 1, 5, 7$, or 11 (see table on page 47). In particular $(-6|5) = 1$.

But it is clear that $5 \neq a^2 + 6b^2$. Similarly, $(-5|3) = +1$, but $3 \neq a^2 + 5b^2$. The partial proof of Theorem 60 using the unique factorization of Gaussian integers (page 150) suggests that the "difficulty" stems from the lack of unique factorization in $k(\sqrt{-5})$ and $k(\sqrt{-6})$ (see page 153). This is indeed the case. The following may be shown.

Theorem 75. *If* $(-6|p) = 1$, $p = a^2 + 6b^2$ *in a unique way if* $p = 24m + 1, 7$. *But* $2p = a^2 + 6b^2$ *if* $p = 24m + 5, 11$. *Similarly, if* $(-5|p) = 1$, $p = a^2 + 5b^2$ *or* $2p = a^2 + 5b^2$ *according as* $p = 20m + 1, 9$ *or* $p = 20m + 3, 7$.

The *two* classes of primes, in either case of this theorem, are related to the so-called *class number* (see page 153), which is >1 when unique factorization is absent. We cannot do justice to this most interesting concept in a few pages. Instead we pass on to other subjects.

Exercise 122. Prove that for $N = 7$ everything is "OK" again—that is, if $(-7|p) = +1$, there is a unique representation $p = a^2 + 7b^2$. The fact that the relatively large value $N = 7$ is still "OK" is related to the specially large density of primes of the form $n^2 + 7$. See table on page 49 and compare remarks about $n^2 + 5, 6$ on page 154.

Exercise 123. For $N = 10$, find a p such that $(-10|p) = +1$, but $p \neq a^2 + 10b^2$.

Exercise 124. In general, if $p < N$, $p \neq a^2 + Nb^2$. What does this suggest concerning unique factorization in $k(\sqrt{-N})$ in general? Investigate the literature to confirm or reject any hypothesis you develop. Caution: If $N \equiv -1 \pmod 4$ the integers of $k(\sqrt{-N})$ are of the form $\frac{1}{2}(a + \sqrt{-N}b)$. By unique factorization one could therefore only conclude that $4p = a^2 + Nb^2$. An example is $p = 3$, $N = 11$. The integers in $k(\sqrt{-11})$ do have unique factorization.

Exercise 125. Analogous to Theorem 68, for $N = 2$ there is the following:

The number of representations of $n = x^2 + 2y^2$ is equal to $2(A - B)$ if n has A divisors $\equiv 1, 3 \pmod 8$, and B divisors $\equiv -1, -3 \pmod 8$.

By an argument similar to that above (page 164) but now using ellipses $x^2 + 2y^2 = n$, show that

$$\frac{\pi}{2\sqrt{2}} = 1 + \frac{1}{3} - \frac{1}{5} - \frac{1}{7} + \frac{1}{9} + \frac{1}{11} - \frac{1}{13} - \frac{1}{15} \cdots.$$

Exercise 126. Conjecture the results analogous to the previous exercise for $N = 3$. Investigate the literature to check your conjecture.

57. Some Applications

We now give several applications of the foregoing results.

(A) (For those who know vector algebra.) Diophantus's formula, Eq. (215), has an interesting interpretation in vector algebra. Let

$$V_1 = ai + bj, \qquad V_2 = ci + dj.$$

Then the scalar and vector products are $V_1 \cdot V_2 = ac + bd$, $V_1 \times V_2 = (ad - bc)\mathbf{k}$. But the magnitude of $V_1 \times V_2$ is the length of V_1 times the length of V_2 times the sine of the angle between them. And $V_1 \cdot V_2$ is the length of V_1 times the length of V_2 times the cosine. Therefore

$$|V_1|^2 |V_2|^2 = (V_1 \cdot V_2)^2 + |V_1 \times V_2|^2,$$

and we obtain the first part of Eq. (215). On the other hand, if $V_3 = ci - di$, while $|V_3| = |V_2|$, the sine and cosine of the angle between V_1 and V_3 will now be different, generally, and we obtain the second representation in Eq. (215).

(B) (For those who know partial differential equations.) If the lowest frequency with which an elastic square membrane can vibrate is

$$\omega_0 = \sqrt{2}k$$

where k is a constant, then it is well known that every possible frequency is given by

$$\omega = \sqrt{a^2 + b^2}k \tag{234}$$

where a and b are positive integers. Corresponding to this frequency, Eq. (234), the shape of the membrane is given by

$$C \sin (\pi a y/L) \sin (\pi b y/L)$$

where L is the length of the side. For the frequency Eq. (234), there will therefore be s different modes of motion if $n = a^2 + b^2$ can be written as a sum of squares in s different ways—where a and b are positive, but where the order is counted. Thus for ω_0, $s = 1$; for $\omega = \sqrt{5}k$, $s = 2$; for $\omega = \sqrt{65}k$, $s = 4$, etc.

(C) (For those who attempted Exercise 16, page 29.) By Theorem 72 the prime $q = 6p + 1$ may be written

$$q = a^2 + 3b^2$$

in a unique way. The criterion sought is this: $q | M_p$ if, and only if, $3|b$.

EXAMPLES:

$$p = 5, \qquad q = 31 = 2^2 + 3 \cdot 3^2; \qquad \text{Since } 3|3,\ 31|M_5$$

$$p = 13, \qquad q = 79 = 2^2 + 3 \cdot 5^2; \qquad \text{Since } 3\nmid 5,\ 79 \nmid M_{13}$$

$$p = 17, \qquad q = 103 = 10^2 + 3 \cdot 1^2; \qquad \text{Since } 3 \nmid 1,\ 103 \nmid M_{17}$$

We shall not prove this rule, but we will indicate its source.

Let g be a primitive root of q, and let $2 \equiv g^e \pmod{q}$, and therefore $2^p \equiv g^{ep} \pmod{q}$. Since $q = 6p + 1$, we have that $q|M_p$ if, and only if, $6|e$. Since $p = 4m + 1$ (see page 29), we have $(2|q) = +1$, and e is even. Therefore $q|M_p$ if, and only if $3|e$. Therefore the necessary and sufficient condition sought is that 2 is a *cubic residue* of q:

$$x^3 \equiv 2 \pmod{q}.$$

Prior to the time that the theory of cubic residues was developed, Gauss found that it was necessary in developing the theory of biquadratic residues, $x^4 \equiv a \pmod{p}$, to introduce the Gaussian integers—namely, those of the algebraic number field $k(e^{2\pi i/4}) = k(i)$. Similarly, under this stimulus, Eisenstein developed the theory of cubic residues with the field $k(e^{2\pi i/3})$. Since

$$e^{2\pi i/3} = \tfrac{1}{2}(-1 + \sqrt{-3}),$$

we are not surprised to find criteria involving

$$a^2 + 3b^2 = (a + \sqrt{-3}\, b)(a - \sqrt{-3}\, b).$$

The criterion that 2 is a cubic residue of $q = 6m + 1$ is: $3|b$, where $q = a^2 + 3b^2$.

(D) (Necessary and Sufficient Conditions for Primality.)

Theorem 76. *For* $n > 1$, *and*

$N = 1$: *assume* $n = 4m + 1$; *for*

$N = 2$: *assume* $n = 8m + 1$ *or* $8m + 3$; *for*

$N = 3$: *assume* $n = 6m + 1$.

If n is prime, $n = a^2 + Nb^2$ in a unique way in positive integers a and b, and $(a, b) = 1$. Conversely, if $n = a^2 + Nb^2$ in a unique way in nonnegative integers, a and b, and if $(a, b) = 1$, then n is prime.

Proof. For n prime we have shown a unique representation. Further $(a, b) = 1$ since $(a, b)|n$.

Now, conversely, let $n = a^2 + Nb^2$ and $(a, b) = 1$. Then $(b, n) = 1$ and $(ab^{-1})^2 \equiv -N \pmod{n}$. Thus every prime divisor of n is of the form listed above corresponding to N. By Theorem 71,

$$\begin{aligned}(a^2 + Nb^2)(c^2 + Nb^2) &= (ac + Nbd)^2 + N(ad - bc)^2 \\ &= (ac - Nbd)^2 + N(ad + bc)^2.\end{aligned} \tag{235}$$

Therefore a product of two primes satisfying $(-N|p) = +1$ is also of the form $x^2 + Ny^2$ with x and y positive. For if $(ac - Nbd)$ and $(ad - bc)$ were both zero, we find $a^2 = Nb^2$. For $N = 2, 3$ this is clearly impossible.

For $N = 1$, likewise—since otherwise $a^2 + b^2$ would be even. Therefore at least one of the representations in Eq. (235) has $x > 0$ and $y > 0$. By induction every divisor of $n > 1$ equals $x^2 + Ny^2$ in positive integers. Therefore if n is composite, write it as a product, Eq. (235), with a, b, c, $d > 0$. Then there are at least two distinct representations of n in non-negative integers, since $ac + Nbd > ac - Nbd$. For $N = 2, 3$ this suffices. For $N = 1$, we must also show that $ac + Nbd = ac + bd \neq ad + bc$. This is so because

$$a(c - d) = b(c - d)$$

implies $c = d$, or $a = b$, and thus that n is even. This completes the proof.

With Theorem 76 we have a method for determining the primality of $n = 4m + 1$ by $N = 1$, and of $n = 8m + 3$ by $N = 2$. The method is useful if n is not too large. One uses subtraction and a table of squares, instead of division and a table of primes. To test the remaining numbers, namely $n = 8m + 7$, one would want to use $N = -2$. But as we have seen in Theorem 74 we now lack uniqueness. To clarify the number of representations of $n = a^2 - 2b^2$ we now investigate Fermat's Equation.

Exercise 127. Show that Theorem 76 may be easily extended to the case $N = -1$ and $n = 2m + 1$.

Exercise 128. $45 = a^2 + b^2$ in a unique way, but 45 is not a prime. $25 = a^2 + b^2$ in a unique way in *positive* integers, but 25 is not prime. $21 \neq a^2 + b^2$, and therefore 21 is composite. Again, 21 is composite since it equals $a^2 + 5b^2$ in two ways. But neither 3 nor 7 equals $a^2 + 5b^2$. From Theorem 75,

$$3 = \tfrac{1}{2}(1^2 + 5 \cdot 1^2), \qquad 7 = \tfrac{1}{2}(3^2 + 5 \cdot 1^2).$$

Thus $3 \cdot 7 = (4 + \sqrt{-5})\,(4 - \sqrt{-5}) = (1 + 2\sqrt{-5})\,(1 - 2\sqrt{-5})$. Compare page 153. Construct a similar example: $pq = a^2 + 6b^2$ in two ways, while neither p nor q equals $a^2 + 6b^2$.

Exercise 129. One half of the numbers $8m + 7$ may be tested by $n = a^2 + 3b^2$.

Exercise 130. All M_p for p an odd prime fall in the class indicated in the previous exercise. In particular M_{11} is not a prime, since $M_{11} \neq a^2 + 3b^2$. But for p large, say $p = 61$, the test is impractical.

58. The Significance of Fermat's Equation

The equation:

$$x^2 - Ny^2 = 1 \tag{236}$$

for $N > 1$, and not a square, is called *Fermat's Equation*. In older writings it is often called "Pell's Equation." If $N = n^2$, it is clear that Eq. (236) has no solution in positive integers since no two positive squares differ by one. Fermat stated that Eq. (236) has infinitely many solutions for every other positive N. He suggested the cases $N = 61$ and 109 as challenge problems. Later Frenicle challenged the English mathematicians with $N = 151$ and 313.

For some N a solution is easily obtained. For $N = 2$ we have $3^2 - N2^2 = 1$ from Theorem 58, and, more generally, if $N = n^2 + 1$,

$$(2n^2 + 1)^2 - N(2n)^2 = 1. \tag{237}$$

But for $N = 61$, $x = 1766319049$ and $y = 226153980$ is the smallest solution, and for $N = 313$ the smallest x has 17 digits. Such an x is not something one would like to obtain by trial and error.

EXERCISE 131. Verify the following generalization of Eq. (237). If $N = (nm)^2 \pm m$, then

$$(2n^2m \pm 1)^2 - N(2n)^2 = 1. \tag{238}$$

And if $N = (nm)^2 \pm 2m$, then

$$(n^2m \pm 1)^2 - N(n)^2 = 1. \tag{239}$$

Show that by a proper choice of m and n, Eqs. (238) and (239) suffice to yield solutions for all nonsquare N where $2 \leqq N \leqq 20$ except for two cases. Likewise for $30 \leqq N \leqq 42$.

In the next section we state and prove the main theorem by a lengthy implicit construction. Later we give an efficient algorithm. We now list some reasons why Eq. (236) is important.

(A) If Eq. (236) is generalized to

$$a^2 - Nb^2 = M \tag{240}$$

for any integer M, there can be no solution unless M is a quadratic residue of every prime which divides N; the example $N = 3$, $M = -1$ was mentioned on page 140. (We note that while this condition is necessary, it is not sufficient. Thus

$$a^2 - 34b^2 = -1$$

has no solution even though -1 is a quadratic residue of 2 and 17.) $M = 1$ is, of course, a quadratic residue of all primes.

(B) But if Eq. (240) has a solution, it has infinitely many. Using the method in the proof of Theorem 74, with the identity from Theorem 71,

$$(x^2 - Ny^2)(a^2 - Nb^2) = (xa \pm Nyb)^2 - N(xb \pm ya)^2, \tag{241}$$

and with any solution of Eq. (236), one obtains another solution of Eq. (240). Further, since we may take $M = 1$, one solution of Eq. (236)

implies infinitely many. All this because $1 \cdot M = M$ on the left side of Eq. (241).

(C) This special role of $M = 1$ is also indicated—it is really the same point in different language—by the fact that for any solution x and y of Eq. (236),

$$x \pm \sqrt{N}\, y$$

is a unity of the algebraic field $k(\sqrt{N})$. See pages 152, 167.

(D) Again, the solutions of Eq. (236) are intimately related to the *rational approximations* of $\sqrt{N}$, as we already noted on pages 139, 140. Thus, from a larger solution for $N = 3$:

$$70226^2 - 3 \cdot 40545^2 = 1,$$

we get

$$70226/40545 = 1.7320508077 \cdots, \tag{242}$$

which agrees with $\sqrt{3}$ to ten figures.

(E) Further, these approximations, and the solutions of Eq. (236), are obtained by infinite *continued fractions,* and Fermat's Equation was the occasion for the introduction of this technique into number theory.

(F) The same continued fractions may be used expeditiously to obtain

$$p = a^2 + b^2$$

for primes of the form $4m + 1$.

(G) If we factor the left side of Eq. (242):

$$\frac{70226}{40545} = \frac{26}{15} \cdot \frac{37}{51} \cdot \frac{73}{53}$$

we obtain convenient *gear ratios* to approximate $\sqrt{3}$:

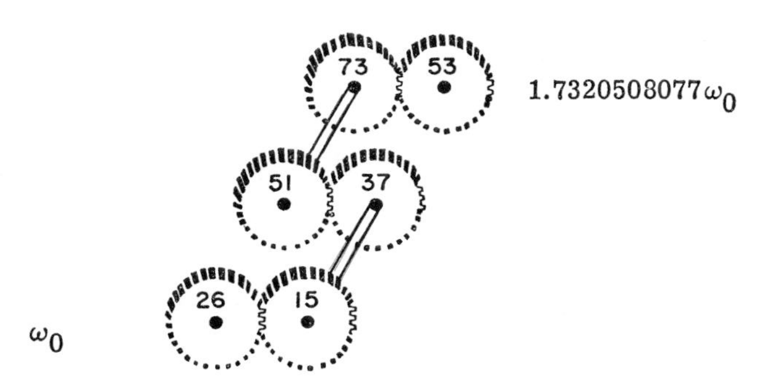

(H) But to carry out such factorizations it is desirable to know the divisibility properties of the solutions (x, y) of Eq. (236). These properties are given by interesting and useful divisibility theorems for the infinite sequence of solutions of Eq. (236). For $N = 3$ these theorems were used by Lucas to obtain his criterion for the primality of Mersenne numbers. It was this consideration (page 120) which led us into this chapter.

59. The Main Theorem

Theorem 77. *If $N > 1$, and not square,*

$$x^2 - Ny^2 = 1 \tag{243}$$

has infinitely many solutions in positive integers. If $x_1 + \sqrt{N}\, y_1$ is the smallest value that $x + \sqrt{N}\, y$ takes on, with x, y a solution, then every solution is given by

$$x_n + \sqrt{N}\, y_n = (x_1 + \sqrt{N}\, y_1)^n. \tag{244}$$

The x_n and y_n may be computed explicitly by

$$\begin{aligned} x_n &= \tfrac{1}{2}[(x_1 + \sqrt{N}\, y_1)^n + (x_1 - \sqrt{N}\, y_1)^n] \\ y_n &= \frac{1}{2\sqrt{N}}[(x_1 + \sqrt{N}\, y_1)^n - (x_1 - \sqrt{N}\, y_1)^n], \end{aligned} \tag{245}$$

or recursively by

$$\begin{aligned} x_{n+1} &= x_1 x_n + N\, y_1 y_n \\ y_{n+1} &= y_1 x_n + x_1 y_n\, . \end{aligned} \tag{246}$$

Comment: If x, y are positive integers which satisfy Eq. (243) we will sometimes use the expression: $x + \sqrt{N}y$ "is" a solution of Eq. (243).

Proof. First we prove that $x_1^2 - Ny_1^2 = 1$ implies $x_n^2 - Ny_n^2 = 1$. From Eq. (241), with $x = a = x_1$ and $y = b = y_1$, and choosing the plus sign on the right, we see that the x_2 and y_2 of Eq. (246), with $n = 1$, satisfy Eq. (243) if x_1 and y_1 do. By induction, the x_n and y_n of Eq. (246) also satisfy Eq. (243). Also, by induction, these integers satisfy Eq. (244), and likewise

$$x_n - \sqrt{N}y_n = (x_1 - \sqrt{N}y_1)^n.$$

Then Eq. (245) follows at once.

Next we prove that there are no other solutions. Assume another solution, Eq. (243). Then

$$x + \sqrt{N}y \neq x_n + \sqrt{N}y_n\, ,$$

for, if equality held, $x - x_n = \sqrt{N}(y_n - y)$, and, since $\sqrt{N}$ is irrational by Theorem 57, we must have $y = y_n$, $x = x_n$. Therefore, since $x_1 +$

$\sqrt{N}y_1 > 1$, and thus, by Eq. (244), $x_n + \sqrt{N}y_n < x_{n+1} + \sqrt{N}y_{n+1}$, assume

$$x_n + \sqrt{N}y_n < x + \sqrt{N}y < x_{n+1} + \sqrt{N}y_{n+1}. \tag{247}$$

Since $(x_n - \sqrt{N}y_n)(x_n + \sqrt{N}y_n) = 1$, we note that the first factor here is >0. Multiply Eq. (247) by this positive number, $x_n - \sqrt{N}y_n$, and we have

$$1 < (x + \sqrt{N}y)(x_n - \sqrt{N}y_n) < (x_1 + \sqrt{N}y_1)^{n+1}(x_1 - \sqrt{N}y_1)^n$$
$$= x_1 + \sqrt{N}y_1.$$

Let

$$(x + \sqrt{N}y)(x_n - \sqrt{N}y_n) = a + \sqrt{N}b$$

where $a = xx_n - Nyy_n$ and $b = yx_n - xy_n$. But

$$a^2 - Nb^2 = (x^2 - Ny^2)(x_n^2 - Ny_n^2) = 1,$$

and since $1 < a + \sqrt{N}b$ we find $0 < a - \sqrt{N}b < 1$. Thus $1 < 2a$ and $0 < 2\sqrt{N}\,b$. Therefore we have a solution of Eq. (243) in positive integers a and b with $a + \sqrt{N}b < x_1 + \sqrt{N}y_1$. Since by the definition of $x_1 + \sqrt{N}y_1$ this cannot be, there is no other solution $x + \sqrt{N}y$.

Finally we come to the *real* problem, to show that Eq. (243) has at least one solution. The first published proof is by Lagrange. Our proof rests on a lemma which we will prove later in two ways.

Lemma. *There is an integer M such that $x^2 - Ny^2 = M$ has infinitely many solutions in positive integers.*

We assume this now and consider M^2 boxes $B_{a,b}$ with $0 \leqq a < |M|$, $0 \leqq b < |M|$. Choose $M^2 + 1$ solutions of $x^2 - Ny^2 = M$ and place each pair (x, y) in the box $B_{a,b}$ if $x \equiv a$, $y \equiv b \pmod{|M|}$. By the Box Principle we therefore have two different solutions:

$$x_1^2 - Ny_1^2 = x_2^2 - Ny_2^2 = M \tag{248}$$

with $x_1 \equiv x_2$, $y_1 \equiv y_2 \pmod{|M|}$. Thus

$$x_1x_2 - Ny_1y_2 \equiv x_1^2 - Ny_1^2 \equiv 0 \equiv x_1y_2 - x_2y_1 \qquad (\text{mod } |M|),$$

and we have $x_1x_2 - Ny_1y_2 = uM$ and $x_1y_2 - x_2y_1 = vM$ with u and v integers.

But, by Eqs. (241) and (248),

$$M^2 = (x_1x_2 - Ny_1y_2)^2 - N(x_1y_2 - x_2y_1)^2 = (u^2 - Nv^2)M^2. \tag{249}$$

Then

$$u^2 - Nv^2 = 1.$$

Now, if $v = 0$, $x_1y_2 = x_2y_1$ and, by Eq. (249),

$$M = \pm(x_1x_2 - Ny_1y_2).$$

Thus

$$Mx_1y_2 = \pm(x_1^2x_2y_2 - Ny_1^2x_2y_2) = \pm Mx_2y_2.$$

Since $x_1 > 0$ and $x_2 > 0$ we have $x_1 = x_2$, and likewise $y_1 = y_2$. Thus $v^2 > 0$ and Eq. (243) has at least one solution $u^2 - Nv^2 = 1$ in positive integers. By the Well-Ordering Principle there is therefore a smallest solution: $x_1 + \sqrt{N}y_1$.

The reader may note that the device used in Eq. (249) of multiplying two equations, and then dividing them by M^2, is analogous to the strategy utilized in Theorem 60, both after Eq. (218) and after Eq. (222).

A proof of the Lemma using a continued fraction algorithm will be given later. A shorter, and now standard, proof runs as follows:

PROOF OF THE LEMMA. For $y = 0, 1, 2, \cdots$, let $x = [\sqrt{N}y] + 1$. Since $\sqrt{N}$ is irrational we have

$$0 < z = x - \sqrt{N}y < 1.$$

For any positive integer n_1 consider the n_1 boxes:

$$0 < z \leqq \frac{1}{n_1}; \frac{1}{n_1} < z \leqq \frac{2}{n_1}; \cdots; \quad \frac{n_1 - 1}{n_1} < z \leqq 1,$$

and $n_1 + 1$ values of z given by $y = 0, 1, \cdots, n_1$. At least two z's are in one box, and they are unequal since $\sqrt{N}$ is irrational. Call them $z_1 > z_2$. Then their difference satisfies:

$$0 < z_1 - z_2 = (x_1 - x_2) - \sqrt{N}(y_1 - y_2) < \frac{1}{n_1}.$$

This may be written

$$0 < z_3 = x_3 - \sqrt{N}y_3 < \frac{1}{n_1} \leqq \frac{1}{|y_3|}$$

where $x_3 = x_1 - x_2$, $y_3 = y_1 - y_2$. Now choose n_2 by

$$\frac{1}{n_2} < z_3$$

and by the same process we obtain a $z_4 < z_3$ with

$$0 < z_4 = x_4 - \sqrt{N}y_4 < \frac{1}{|y_4|}.$$

Thus, since $\sqrt{N}$ is irrational, there are infinitely many solutions of

$$0 < x - \sqrt{N}y < \frac{1}{|y|}. \tag{250}$$

It follows that

$$2\sqrt{N}y < x + \sqrt{N}y < 2\sqrt{N}y + 1,$$

and thus, whether y and x are positive or negative,

$$|x + \sqrt{N}y| < 2\sqrt{N}|y| + 1.$$

Therefore we find infinitely many solutions of

$$0 < |x^2 - Ny^2| = (x - \sqrt{N}y)|x + \sqrt{N}y| < 2\sqrt{N} + 1.$$

By the Box Principle (extended) we therefore have infinitely many solutions of $x^2 - Ny^2 = M$ for some $0 < |M| < 2\sqrt{N} + 1$. This completes the proof of Theorem 77.

The reader notes the curious character of the proof given here for Theorem 77. A solution of Eq. (243) implies

$$\frac{x}{y} - \sqrt{N} = \frac{1}{y(x + \sqrt{N}y)},$$

that is, x/y is a "good" rational approximation of $\sqrt{N}$. In the proof of the Lemma, we first find that there exist approximations:

$$\frac{x}{y} - \sqrt{N} < \frac{1}{y};$$

then, by the Box Principle, better approximations:

$$\left|\frac{x}{y} - \sqrt{N}\right| < \frac{1}{y^2}.$$

Finally, using the Lemma, we attain the required approximations. It could be called a proof by "convergence," and this suggests that an *explicit* and more efficient algorithm for finding good rational approximations of $\sqrt{N}$ could lead to an explicit and more efficient construction of solutions of Eq. (243). This we now examine.

EXERCISE 132. By the same technique as that used on page 174 to show that there is no other solution $x + \sqrt{N}y$, show that if $u^2 - Ny^2 = -1$ has a solution, and if $u_1 + \sqrt{N}y_1$ is the smallest value possible, then all solutions of $u^2 - Ny^2 = -1$ are given by

$$u_n + \sqrt{N}y_n = (u_1 + \sqrt{N}y_1)^n \tag{251}$$

for *n odd*, while for *n even* one obtains the solutions of Fermat's Equation:

$$u_{2m} + \sqrt{N}v_{2m} = x_m + \sqrt{N}y_m \tag{252}$$

given by Eq. (244).

EXERCISE 133. Show that Theon's rule, Eq. (171), gives all solutions of

$$x^2 - 2y^2 = \pm 1.$$

60. AN ALGORITHM

For any positive nonsquare N we define five sequences of nonnegative integers A_n, B_n, C_n, P_n, and Q_n as follows. Let

$$C_{-1} = N; \quad C_0 = 1; \quad B_0 = 0; \quad P_{-1} = Q_0 = 0; \quad P_0 = Q_{-1} = 1. \tag{253}$$

For $n = 0, 1, 2, \cdots$, define the sequences recursively by

$$A_{n+1} = \left[\frac{\sqrt{N} + B_n}{C_n}\right], \quad \text{or,} \quad \text{since } A_1 = [\sqrt{N}],$$

$$A_{n+1} = \left[\frac{A_1 + B_n}{C_n}\right]. \tag{254}$$

$$B_{n+1} = A_{n+1}C_n - B_n. \tag{255}$$

$$C_{n+1} = C_{n-1} + A_{n+1}(B_n - B_{n+1}). \tag{256}$$

$$P_{n+1} = P_{n-1} + A_{n+1}P_n. \tag{257}$$

$$Q_{n+1} = Q_{n-1} + A_{n+1}Q_n. \tag{258}$$

In Eq. (254) we use the [ ] function of page 14.

EXAMPLE:
For $N = 19$ we show the sequences:

n	C_n	A_n	B_n	P_n	Q_n
−1	19	—	—	0	1
0	1	—	0	1	0
1	3	4	4	4	1
2	5	2	2	9	2
3	2	1	3	13	3
4	5	3	3	48	11
5	3	1	2	61	14
6	1	2	4	170	39
7	3	8	4	1421	326

It is clear, by the rules (Eqs. 254–256), that since C_{n-1}, B_n and C_n repeat here for $n = 1$ and 7, that A_n, B_n, and C_n will henceforth be periodic with a *period* of 6.

Our immediate interest in the algorithm—there will be other points later—is in the important relation which we will prove in the following section:

$$P_n^2 - NQ_n^2 = (-1)^n C_n. \tag{259}$$

If $C_n = 1$, with n even, we obtain a solution of Fermat's Equation: $P_n^2 - NQ_n^2 = 1$. It will be shown that for every N there are infinitely many n with $(-1)^n C_n = 1$, and for the smallest such $n > 0$ we obtain the smallest solution of Eq. (243):

$$P_n + \sqrt{N}Q_n = x_1 + \sqrt{N}y_1.$$

Thus $170 + \sqrt{19}\cdot 39$ is the smallest solution for $N = 19$.

If $C_n = 1$ with n odd, the smallest such n yields the smallest solution of $u^2 - Nv^2 = -1$ by

$$P_n + \sqrt{N}Q_n = u_1 + \sqrt{N}v_1$$

using the notation of Exercise 132. We show such a case for $N = 13$:

n	C_n	A_n	B_n	P_n	Q_n
−1	13	—	—	0	1
0	1	—	0	1	0
1	4	3	3	3	1
2	3	1	1	4	1
3	3	1	2	7	2
4	4	1	1	11	3
5	1	1	3	18	5
6	4	6	3	119	33

Then $18 + \sqrt{13}\cdot 5$ is the smallest solution of $u^2 - 13v^2 = -1$. Correspondingly, as in Exercise 132, $649 + \sqrt{13}\cdot 180 = (18 + \sqrt{13}\cdot 5)^2$ is the smallest solution of $x^2 - 13y^2 = 1$. Alternatively, one could continue the table until $C_{10} = C_5 = 1$, since now there is a *period* of 5. Then $P_{10} = 649$, $Q_{10} = 180$. In fact, by periodicity we have

$$P_{5k}^2 - 13Q_{5k}^2 = (-1)^k.$$

Exercise 134. Obtain solutions of $x^2 - 31y^2 = 1$, and of $x^2 - 41y^2 = \pm 1$.

EXERCISE 135. For $N = (nm)^2 + m$ (compare Exercise 131) carry out the algorithm *algebraicly* and obtain

$$P_2 = 2n^2m + 1, \qquad Q_2 = 2n.$$

Similarly carry out the algorithm for $N = (nm)^2 - m$. In this case what is the period of A_n, B_n, C_n if $m = 1$; if $m > 1$?

We shall see that A_n, B_n, and C_n are periodic, from some point on, for all N. This will follow from the inequalities:

$$0 < A_n < 2\sqrt{N}, \qquad 0 < B_n < \sqrt{N}, \qquad 0 < C_n < 2\sqrt{N} \qquad (260)$$

which hold for all positive n. Granting these for now, it is clear, by the Box Principle, that C_{n-1}, B_n, and C_n must eventually repeat. Then A_n, B_n, and C_n will be periodic henceforth. We designate the period $\rho(N)$, as in $\rho(19) = 6$, $\rho(13) = 5$. Assuming Eq. (259), we then have another proof of the Lemma (page 175).

We will also obtain the useful invariant:

$$B_n^2 + C_nC_{n-1} = N. \qquad (261)$$

Since $N \neq m^2$, we see that $C_n \neq 0$. This justifies the division in Eq. (254). Again, since $C_{n-1} = (N - B_n^2)/C_n$ we need not have stipulated the repetition of C_{n-1} in the previous paragraph. Also, if $C_n = C_{n-1}$ we have $N = B_n^2 + C_n^2$. In particular, for $n = 3$, we have $13 = 2^2 + 3^2$. It can be shown that for every prime $N = 4m + 1$ there is such a $C_n = C_{n-1}$.

61. CONTINUED FRACTIONS FOR $\sqrt{N}$

Consider

$$(\sqrt{2} - 1)(\sqrt{2} + 1) = 1.$$

Then

$$\sqrt{2} = 1 + \frac{1}{1 + \sqrt{2}}$$

and, by substitution we have

$$\sqrt{2} = 1 + \frac{1}{2+}\,\frac{1}{1 + \sqrt{2}}.$$

By continuation we obtain the infinite continued fraction:

$$\sqrt{2} = 1 + \frac{1}{2+}\,\frac{1}{2+}\,\frac{1}{2+}\cdots$$

which we abbreviate as

$$\sqrt{2} = 1 + \frac{1}{2+}\frac{1}{2+}\frac{1}{2+}\cdots. \tag{262}$$

The fractions

$$\frac{1}{1} = 1, \quad \frac{3}{2} = 1 + \frac{1}{2}, \quad \frac{7}{5} = 1 + \frac{1}{2+\frac{1}{2}} = 1 + \frac{1}{2+}\frac{1}{2}, \quad \text{etc.}$$

are called the *convergents* of the continued fraction. The reader may note that these convergents are c_n/a_n, the ratios of Theon's diagonal and side numbers.

If $\sqrt{3} = 1 + (1/x)$, we have $x = 1/(\sqrt{3} - 1) = (\sqrt{3} + 1)/2$. Let $(\sqrt{3} + 1)/2 = 1 + (1/y)$ and $y = 2/(\sqrt{3} - 1) = \sqrt{3} + 1 = 2 + (1/x)$. Thus

$$\sqrt{3} = 1 + \frac{1}{1+}\frac{1}{2+}\frac{1}{1+}\frac{1}{2+}\cdots \tag{263}$$

The reader may verify that the convergents now:

$$\frac{c}{a} = \frac{1}{1}, \quad \frac{2}{1}, \quad \frac{5}{3}, \quad \frac{7}{4}, \quad \text{etc.}$$

are alternately solutions of Eqs. (176) and (175), and, conversely, Archimedes' approximations (page 140) are later convergents.

It may be easily shown that the convergents of any continued fraction

$$A_1 + \frac{1}{A_2+}\frac{1}{A_3+}\frac{1}{A_4+}\cdots$$

form a convergent sequence if the A's are positive integers. Also that if x is irrational and >1 it has a *unique* representation of this type:

$$x = A_1 + \frac{1}{A_2+}\frac{1}{A_3+}\cdots$$

Further, if such an x is given by

$$x = A_1 + \frac{1}{A_2+}\frac{1}{A_3+}\cdots\frac{1}{+A_n+}\frac{1}{y} \tag{264}$$

where $y > 1$, then these n values of A are those of its unique representation. It follows that Eqs. (262) and (263) are the representations for $\sqrt{2}$ and $\sqrt{3}$.

One could proceed with $\sqrt{N}$ as with $\sqrt{3}$. But there is much redundancy, notationwise and otherwise, in such algebra. If one seeks an algorithm with the redundancy removed one obtains that given in Sect. 60. Thus we shall see that

$$\sqrt{N} = A_1 + \frac{1}{A_2 +}\frac{1}{A_3 +}\frac{1}{A_4 +} \cdots \tag{265}$$

where the A's are given by Eq. (254). Further the convergents are

$$\frac{P_n}{Q_n} = A_1 + \frac{1}{A_2 +}\frac{1}{A_3 +} \cdots \frac{1}{+ A_n} \tag{266}$$

where P_n and Q_n are given by Eqs. (257–258). Thus, from page 178,

$$\sqrt{19} = 4 + \frac{1}{2 +}\frac{1}{1 +}\frac{1}{3 +}\frac{1}{1 +}\frac{1}{2 +}\frac{1}{8 +}\frac{1}{2 +} \cdots.$$

We may indicate the periodicity neatly by the symmetric formula:

$$\sqrt{19} = 4 + \frac{1}{2 +}\frac{1}{1 +}\frac{1}{3 +}\frac{1}{1 +}\frac{1}{2 +}\frac{1}{4 + \sqrt{19}}.$$

Similarly

$$\sqrt{13} = 3 + \frac{1}{1 +}\frac{1}{1 +}\frac{1}{1 +}\frac{1}{1 +}\frac{1}{3 + \sqrt{13}}$$

and

$$\frac{18}{5} = 3 + \frac{1}{1 +}\frac{1}{1 +}\frac{1}{1 +}\frac{1}{1}.$$

We now prove Eqs. (259), (260), (261), (265) and (266). The subject of continued fractions is a large one. It is not our purpose now to expound upon it at length. Our primary interest concerns its relation to Theorem 77. At that, our treatment is brief and we leave numerous computations for the reader.

First, from Eqs. (255) and (256) we have

$$C_{n+1} = C_{n-1} + 2A_{n+1}B_n - A_{n+1}^2C_n,$$

and from this, and Eq. (255), we obtain Eq. (261) by induction. Then we define

$$\alpha_n = \frac{\sqrt{N} + B_{n-1}}{C_{n-1}}, \tag{267}$$

and using Eq. (261) we find that

$$\alpha_n = A_n + \frac{1}{\alpha_{n+1}}. \tag{268}$$

Since $\alpha_1 = \sqrt{N}$ we obtain

$$\sqrt{N} = A_1 + \frac{1}{A_2 +}\,\frac{1}{A_3 +}\cdots+\frac{1}{A_n +}\,\frac{1}{\alpha_{n+1}} \tag{269}$$

by induction.

Next we show $0 < A_n$ and $1 < \alpha_n$. From $\alpha_n > 1$, and $[\alpha_n] = A_n$, and Eq. (268), we find $A_n > 0$ and $\alpha_{n+1} > 1$. Since $\alpha_1 = \sqrt{N} > 1$, the required results follow by induction. Then from Eqs. (269) and (264) we derive Eq. (265).

To complete the proof of Eq. (260) assume

$$0 < B_n < \sqrt{N}, \qquad 0 < C_n < 2\sqrt{N} \tag{270}$$

for some positive n. From

$$\alpha_{n+1} - A_{n+1} = \frac{\sqrt{N} - B_{n+1}}{C_n} > 0$$

we find $B_{n+1} < \sqrt{N}$. Then from $B_{n+1}^2 + C_{n+1}C_n = N$, we obtain $0 < C_{n+1}$. Thus from

$$\alpha_{n+2} = \frac{\sqrt{N} + B_{n+1}}{C_{n+1}} > 1$$

we have $C_{n+1} < 2\sqrt{N}$. But if $B_{n+1} \leqq 0$, from Eq. (255) $C_n \leqq A_{n+1}C_n \leqq B_n < \sqrt{N}$. This implies

$$\alpha_{n+1} - A_{n+1} = \frac{\sqrt{N} - B_{n+1}}{C_n} > 1\,,$$

and this contradiction implies $0 < B_{n+1}$. Then Eq. (270) follows by induction for all $n > 0$. Finally, since $1 \leqq C_n$, we get $A_n < \alpha_n < 2\sqrt{N}$ from Eq. (267). This completes the proof of Eq. (260).

Next, from Eqs. (257), (258) and (253) we obtain a second important invariant by induction:

$$(-1)^n(P_nQ_{n-1} - P_{n-1}Q_n) = 1. \tag{271}$$

This implies that $(P_n, Q_n) = 1$ and the fraction P_n/Q_n is in its lowest terms.

Now we prove Eq. (266), slightly generalized. Let a_n, $n = 1, 2, \cdots$ be any positive numbers, not necessarily integers, and let

$$p_{-1} = q_0 = 0, \qquad p_0 = q_{-1} = 1,$$

$$p_{n+1} = p_{n-1} + a_{n+1}p_n\,, \qquad q_{n+1} = q_{n-1} + a_{n+1}q_n\,,$$

analogous to Eqs. (253), (257) and (258). Then

$$\frac{p_1}{q_1} = \frac{a_1}{1} \quad \text{and} \quad \frac{p_2}{q_2} = \frac{a_2\, a_1 + 1}{a_2} = a_1 + \frac{1}{a_2}$$

are identities. Assume, for some $n > 1$,

$$\frac{p_n}{q_n} = \frac{p_{n-2} + a_n\, p_{n-1}}{q_{n-2} + a_n\, q_{n-1}} = a_1 + \frac{1}{a_2 +}\,\frac{1}{a_3 +} \cdots \frac{1}{+\, a_n} \tag{272}$$

for *any* positive a's. Thus we may replace a_n by $a_n + (1/a_{n+1})$ and obtain

$$\frac{p_{n-2} + \left(a_n + \dfrac{1}{a_{n+1}}\right) p_{n-1}}{q_{n-2} + \left(a_n + \dfrac{1}{a_{n+1}}\right) q_{n-1}} = \frac{p_n + \dfrac{1}{a_{n+1}}\, p_{n-1}}{q_n + \dfrac{1}{a_{n+1}}\, q_{n-1}} = \frac{p_{n+1}}{q_{n+1}}$$

$$= a_1 + \frac{1}{a_2 +} \cdots \frac{1}{+\, a_n +}\,\frac{1}{a_{n+1}}.$$

Therefore Eq. (272) is true for all positive n and a's. In particular, Eq. (266) is true. Further, from Eq. (269), we obtain

$$\sqrt{N} = \frac{P_{n-2} + \left(A_n + \dfrac{1}{\alpha_{n+1}}\right) P_{n-1}}{Q_{n-2} + \left(A_n + \dfrac{1}{\alpha_{n+1}}\right) Q_{n-1}} = \frac{P_{n-1} + \alpha_{n+1}\, P_n}{Q_{n-1} + \alpha_{n+1}\, Q_n}, \tag{273}$$

and, from Eq. (267),

$$\sqrt{N}(P_n - C_n Q_{n-1} - B_n Q_n) = N Q_n - C_n P_{n-1} - B_n P_n.$$

But $\sqrt{N}$ is irrational, so:

$$\begin{aligned} P_n &= C_n Q_{n-1} + B_n Q_n\,, \\ N Q_n &= C_n P_{n-1} + B_n P_n\,. \end{aligned} \tag{274}$$

These combine to yield

$$P_n^{\,2} - N Q_n^{\,2} = C_n(P_n Q_{n-1} - P_{n-1} Q_n),$$

and from Eq. (271) we prove Eq. (259).

It is easy to show that the right side of Eq. (265) converges to the left side, for from Eqs. (273) and (271) we also obtain

$$\frac{P_n}{Q_n} - \sqrt{N} = (-1)^n \frac{1}{Q_n(Q_{n-1} + \alpha_{n+1}\, Q_n)}. \tag{275}$$

Since Q_n increases without bound the convergents converge to $\sqrt{N}$ in an alternating manner:

$$\frac{P_{2k}}{Q_{2k}} > \sqrt{N} > \frac{P_{2k+1}}{Q_{2k+1}}.$$

We have shown that the algorithm yields a convergent, periodic, continued fraction for $\sqrt{N}$, and, if $(-1)^n C_n = 1$, we have a solution of Eq. (243). These fractions were used by Fermat, Frenicle, Wallis and Brouncker to obtain solutions. No one prior to Lagrange, however, (except possibly Fermat), proved that such an n always existed. We have seen that the algorithm implies the Lemma, and this implies a solution: $x^2 - Ny^2 = 1$. Therefore

$$\frac{x}{y} = \sqrt{N + \frac{1}{y^2}},$$

and

$$\frac{x}{y} - \sqrt{N} = \frac{1}{y^2} \frac{1}{\sqrt{N} + \sqrt{N + \frac{1}{y^2}}}. \tag{276}$$

Now any rational number $b/a > 1$ may be expanded into a finite continued fraction as on page 12. We have

$$\frac{b}{a} = q_0 + \frac{1}{q_1 +} \frac{1}{q_2 +} \cdots \frac{1}{+ q_n}$$

where the q's are given by Euclid's Algorithm on page 9. Further $q_n > 1$, and, at our option, we may also write

$$\frac{b}{a} = q_0 + \frac{1}{q_1 +} \cdots \frac{1}{+ (q_n - 1) +} \frac{1}{1}.$$

Using one or the other, x/y can be written

$$\frac{x}{y} = a_1 + \frac{1}{a_2 +} \cdots \frac{1}{a_n} \tag{277}$$

with n *even*. If z is defined by

$$\sqrt{N} = a_1 + \frac{1}{a_2 +} \cdots \frac{1}{+ a_n +} \frac{1}{z},$$

we have, analogous to Eq. (275),

$$\frac{x}{y} - \sqrt{N} = \frac{1}{y(y' + zy)} = \frac{1}{y^2} \cdot \frac{1}{z + \frac{y'}{y}} \tag{278}$$

where y' is the denominator of the next to the last convergent of Eq. (277). Therefore $0 < y'/y < 1$, and comparing Eqs. (278) and (276) we find $z > 1$. Then, by Eq. (264), $a_i = A_i$ and x/y is a convergent P_n/Q_n. It follows that every solution of Eq. (243) is given by the algorithm.

EXERCISE 136. Solve $61 = a^2 + b^2$ and $x^2 - 61y^2 = -1$ by the algorithm. Solve $x^2 - 61y^2 = +1$; compare page 172. Obtain the representation of $\sqrt{61}$.

EXERCISE 137. Let n be the smallest positive index for which $C_n = 1$. From Eqs. (261), (260), etc. show $B_n = B_{n+1} = A_1$ and $A_{n+1} = 2A_1$. The representation may be written

$$\sqrt{N} = A_1 + \frac{1}{A_2 +}\,\frac{1}{A_3 +}\cdots\frac{1}{+ A_n +}\,\frac{1}{A_1 + \sqrt{N}}, \tag{279}$$

and the *period* $\rho(N) = n$. The sequence of A's is

$$A_1,\ A_2,\ A_3 \cdots A_n, \qquad 2A_1,\ A_2,\ A_3, \cdots A_n, \qquad 2A_1, \text{ etc.}$$

EXERCISE 138. The representation, Eq. (279), is always *symmetric*:

$$\sqrt{N} = A_1 + \frac{1}{A_2 +}\,\frac{1}{A_3 +}\cdots\frac{1}{+ A_3 +}\,\frac{1}{A_2 +}\,\frac{1}{A_1 + \sqrt{N}}. \tag{280}$$

To prove Eq. (280), show that one may replace $\sqrt{N}$ by $-\sqrt{N}$ in Eq. (279). Then solve for the lower radical in terms of the upper. Alternatively, use Eqs. (279) and (273) to derive

$$P_n = A_1Q_n + Q_{n-1}.$$

Then, with Eq. (282) below, show that

$$\frac{P_n}{Q_n} = A_1 + \frac{1}{A_n +}\,\frac{1}{A_{n-1} +}\cdots\frac{1}{+ A_2}.$$

EXERCISE 139. Show that if one runs the indices backwards one obtains

$$A_n + \frac{1}{A_{n-1} +}\,\frac{1}{A_{n-2} +}\cdots\frac{1}{+ A_1} = \frac{P_n}{P_{n-1}} \tag{281}$$

and

$$A_n + \frac{1}{A_{n-1} +}\cdots\frac{1}{+ A_2} = \frac{Q_n}{Q_{n-1}}. \tag{282}$$

EXERCISE 140.

$$A_1 + \frac{1}{A_2 +}\cdots\frac{1}{+ A_m +}\,\frac{1}{A_{m+1} +}\,\frac{1}{A_m +}\,\frac{1}{A_{m-1} +}\cdots\frac{1}{+ A_2} = \frac{P_{m+1}Q_m + P_mQ_{m-1}}{Q_m(Q_{m-1} + Q_{m+1})}. \tag{283}$$

$$A_1 + \frac{1}{A_2 +} \cdots + \frac{1}{A_m +} \frac{1}{A_m +} \cdots + \frac{1}{A_2} = \frac{P_m Q_m + P_{m-1} Q_{m-1}}{Q_m^2 + Q_{m-1}^2}. \tag{284}$$

EXERCISE 141. Use one of the results of the previous exercise as a short-cut in solving $x^2 - 61y^2 = -1$. What do you note about the P's and Q's used, in relation to the $61 = a^2 + b^2$ of Exercise 136?

EXERCISE 142. From Eq. (273) and the periodicity of the A's rederive the recurrence relations, Eq. (246).

EXERCISE 143. There are infinitely many solutions of $x^2 - 34y^2 = M$ for $M = +2$ and -9, but none for $M = -1$.

EXERCISE 144. If $N \equiv 1 \pmod 4$ and prime, and the smallest solution of Eq. (243) is $x_1 + \sqrt{N}y_1$, then

$$u_1 = \sqrt{\frac{x_1 - 1}{2}} \quad \text{and} \quad v_1 = \sqrt{\frac{x_1 + 1}{2N}} \tag{285}$$

are integers and $u_1 + \sqrt{N}v_1$ is the smallest solution of $u^2 - Nv^2 = -1$.

EXERCISE 145. If $N = 2k + 1$ is prime, the period $\rho(N)$ of $\sqrt{N}$ is even or odd according as k is odd or even.

EXERCISE 146. If $N \equiv 1 \pmod 4$ and prime, and if $\rho(N) = 2k - 1$, then

$$N = B_k^2 + C_k^2. \tag{286}$$

EXERCISE 147. If N is an odd prime, and $N | s^2 + 1$ with $0 < s < \frac{1}{2}N$, write

$$\frac{N}{s} = a_1 + \frac{1}{a_2 +} \cdots + \frac{1}{a_{2n}}$$

by Euclid's Algorithm. Then the a's are symmetric and

$$N = p_{n-1}^2 + p_n^2. \tag{287}$$

For example: $1429 | 620^2 + 1$.

$$\frac{1429}{620} = 2 + \frac{1}{3 +} \frac{1}{3 +} \frac{1}{1 +} \frac{1}{1 +} \frac{1}{3 +} \frac{1}{3 +} \frac{1}{2}.$$

$$\frac{p_3}{q_3} = \frac{23}{10}, \quad \frac{p_4}{q_4} = \frac{30}{13}, \quad 1429 = 23^2 + 30^2$$

EXERCISE 148. Conversely, if an odd prime N is a sum of two squares, consider its representation by Eq. (287). Then expand by Euclid's Algorithm:

$$\frac{p_n}{p_{n-1}} = a_{n+1} + \frac{1}{a_{n+2} +} \cdots + \frac{1}{a_{2n}}.$$

If the next to the last convergent is u/v and $s = up_n + vp_{n-1}$, then $0 < s < N$, and $N | s^2 + 1$.

EXERCISE 149. In Exercise 147 it is not necessary to complete Euclid's Algorithm in order to determine n. The largest numerator $< \sqrt{N}$ is p_n.

EXERCISE 150.

$$\rho(N) < 2N. \tag{288}$$

EXERCISE 151.

$$\text{If } \beta_n = \frac{P_n}{Q_n} \text{ then } \beta_n = \frac{N + B_n \beta_{n-1}}{B_n + \beta_{n-1}}. \tag{289}$$

EXERCISE 152. If $x_n + \sqrt{N} y_n$ is the nth solution of Eq. (243), then the $2n$th solution is given by Newton's Algorithm for taking square roots:

$$x_{2n}/y_{2n} = \frac{1}{2}\left[x_n/y_n + \frac{N}{x_n/y_n}\right] \tag{290}$$

if the right side is in its lowest terms.

62. FROM ARCHIMEDES TO LUCAS

From

$$\sqrt{3} = 1 + \frac{1}{1} + \frac{1}{2} + \frac{1}{1} + \frac{1}{2} + \cdots$$

we obtain the approximations P_n/Q_n:

n	1	2	3	4	5	6
$\frac{P_n}{Q_n}$	$\frac{1}{1}$	$\frac{2}{1}$	$\frac{5}{3}$	$\frac{7}{4}$	$\frac{19}{11}$	$\frac{26}{15}$
n	7	8	9	10	11	12
$\frac{P_n}{Q_n}$	$\frac{71}{41}$	$\frac{97}{56}$	$\frac{265}{153}$	$\frac{362}{209}$	$\frac{989}{571}$	$\frac{1351}{780}$

Archimedes' approximations on page 140 are those for $n = 12$ and 9. Our gear ratio on page 173 is

$$\frac{P_{18}}{Q_{18}} = \frac{70226}{40545} = \frac{26}{15} \cdot \frac{37}{51} \cdot \frac{73}{53},$$

and we note that the first factor on the right is P_6/Q_6. This is not an isolated result, for we shall prove

Theorem 78. *For all positive n, r, and s,*

$$Q_n | Q_{rn}\,, \qquad P_n | P_{(2s+1)n}\,. \tag{291}$$

It will be convenient in such investigations to introduce two new sequences. From Theorem 77 we obtain

$$P_{2n} \pm \sqrt{3}Q_{2n} = (P_2 \pm \sqrt{3}Q_2)^n = (2 \pm \sqrt{3})^n. \tag{292}$$

This implies

$$P_{2n+2} = 2P_{2n} + 3Q_{2n}\,, \qquad Q_{2n+2} = P_{2n} + 2Q_{2n}\,. \tag{293}$$

But since $A_{2n+2} = 1$ for all n, we also have, from Eqs. (257) and (258),

$$P_{2n+2} = P_{2n} + P_{2n+1}\,, \qquad Q_{2n+2} = Q_{2n} + Q_{2n+1}\,.$$

Then, from Eq. (293), we obtain the odd-order convergents:

$$P_{2n+1} = P_{2n} + 3Q_{2n}\,, \qquad Q_{2n+1} = P_{2n} + Q_{2n}$$

or

$$P_{2n+1} \pm \sqrt{3}Q_{2n+1} = (P_{2n} \pm \sqrt{3}Q_{2n})(1 \pm \sqrt{3}) = (2 \pm \sqrt{3})^n(1 \pm \sqrt{3}). \tag{294}$$

Now $(1 \pm \sqrt{3})^2 = 2(2 \pm \sqrt{3})$, and Eqs. (292) and (294) may be written

$$2^n(P_{2n} \pm \sqrt{3}Q_{2n}) = (1 \pm \sqrt{3})^{2n},$$
$$2^n(P_{2n+1} \pm \sqrt{3}Q_{2n+1}) = (1 \pm \sqrt{3})^{2n+1}.$$

Using the square bracket $\left[\frac{2n}{2}\right] = \left[\frac{2n+1}{2}\right] = n$, we therefore have, for all m,

$$2^{\left[\frac{m}{2}\right]}(P_m \pm \sqrt{3}Q_m) = (1 \pm \sqrt{3})^m. \tag{295}$$

If we now define

$$2^{\left[\frac{n}{2}\right]}P_n = t_n\,, \qquad 2^{\left[\frac{n}{2}\right]}Q_n = u_n\,, \tag{296}$$

we have

$$t_n \pm \sqrt{3}u_n = (1 \pm \sqrt{3})^n. \tag{297}$$

By this definition we override the pulsing character of $P_n + \sqrt{3}Q_n$—due to the period, $\rho(3) = 2$—and may transfer our investigation to the smooth sequence $t_n + \sqrt{3}u_n$ instead. For, if we can factor t_n and u_n, we can also factor P_n and Q_n by Eq. (296).

From Eq. (297) we obtain at once some useful identities:

$$\begin{aligned} t_{n+m} &= t_n t_m + 3u_n u_m , \\ u_{n+m} &= u_n t_m + u_m t_n . \end{aligned} \tag{298}$$

Then

$$\begin{aligned} t_{2n} &= t_n^2 + 3u_n^2, \\ u_{2n} &= 2u_n t_n . \end{aligned} \tag{299}$$

Since $(1 + \sqrt{3})^n(1 - \sqrt{3})^n = (-2)^n$, we have

$$t_m - \sqrt{3}u_m = (1 - \sqrt{3})^m = (-2)^m(1 + \sqrt{3})^{-m}.$$

Therefore

$$(t_m - \sqrt{3}u_m)(t_n + \sqrt{3}u_n) = (-2)^m(1 + \sqrt{3})^{n-m},$$

and, if $m < n$,

$$\begin{aligned} (-2)^m u_{n-m} &= u_n t_m - u_m t_n , \\ (-2)^m t_{n-m} &= t_n t_m - 3u_n u_m , \end{aligned} \tag{300}$$

while, if $m = n$,

$$(-2)^n = t_n^2 - 3u_n^2. \tag{301}$$

Now we can give the

Proof of Theorem 78. From Eq. (298), if $m = rn$, we see that $u_n|u_{rn}$ implies $u_n|u_{(r+1)n}$. By induction, $u_n|u_{rn}$ for all positive r. Now $t_n|u_{2n}$ by Eq. (299), and therefore, by what has just been proven, $t_n|u_{2sn}$. Thus, if $m = 2sn$, we see from Eq. (298) that $t_n|t_{(2s+1)n}$. Then Eq. (291) follows from Eq. (296) directly if Q_n, or P_n, respectively, is odd. To determine their divisibility by powers of 2, we obtain from Eq. (298), with $m = 4$, and from Eq. (296),

$$\begin{aligned} P_{n+4} &= 7P_n + 12Q_n \\ Q_{n+4} &= 7Q_n + 4P_n . \end{aligned}$$

It follows, by induction, that P_{4k+2} is divisible by 2, but not by 4, and all other P_n are odd. Likewise Q_n is even only for $n = 4k$. From Eqs. (299) and (296) we have

$$Q_{4k} = 2P_{2k}Q_{2k} .$$

It follows, by induction, that for $k > 1$ and $n = 2^k(2s + 1)$, Q_n is divisible by 2^k but not by 2^{k+1}.

Thus Eq. (291) is true for all n.

Corollary. *If Q_n is a prime, then n is a prime.*

We find that Q_3, Q_5, Q_7, Q_{11}, and $Q_{13} = 2131$ are indeed primes. But $Q_{17} = 67 \cdot 443$. $Q_{19} = 110771$ is again prime. This corollary, the numerical behavior (6 primes and one composite), and the exponential growth of the Q_n are all reminiscent of $M_n = 2^n - 1$. Since, from Eqs. (296) and (297), we have

$$Q_n = \frac{(1+\sqrt{3})^n - (1-\sqrt{3})^n}{2\sqrt{3}\cdot 2^{[n/2]}} \tag{302}$$

and see that their formulas are somewhat similar. Let us pursue this analogy.

From Theorem 35, on page 72, if $m > 1$ and odd, and if $m|2^f - 1$, and if e is the smallest positive x such that $m|2^x - 1$, then $e|f$. The analogous result is

Theorem 79. *If $m > 1$ and odd, and if $m|Q_f$, and if Q_e is the smallest positive Q_n which m divides, then $e|f$.*

Proof. Assume the contrary, and let $f = qe + r$ with $0 < r < e$. Consider Eq. (300) with $n = f$, $m = r$. Then

$$(-2)^r u_{qe} = u_f t_r - u_r t_f .$$

Since m divides Q_e, it divides Q_{qe} and therefore u_{qe}. Likewise $m|u_f$, and thus $m|u_r t_f$. Then, since $(t_f, m) = 1$ by Eq. (301), $m|u_r$, and $m|Q_r$. Since this contradicts the definition of e, we have $r = 0$ and $e|f$.

Now we investigate the analogue of *Fermat's Theorem*. Let p be an odd prime, and, using the binomial theorem, we expand

$$(1 \pm \sqrt{3})^p = 1 \pm \frac{p}{1}\sqrt{3} + \frac{p(p-1)}{1\cdot 2}3 \pm \cdots + \frac{p(p-1)\cdots 2}{1\cdot 2\cdots(p-1)}3^{(p-1)/2} \pm 3^{p/2}.$$

Then

$$t_p = \frac{1}{2}[(1+\sqrt{3})^p + (1-\sqrt{3})^p]$$

$$= 1 + \frac{p(p-1)}{1\cdot 2}3 + \cdots + \frac{p(p-1)\cdots 2}{1\cdot 2\cdots(p-1)}3^{(p-1)/2}.$$

But every term except the first is divisible by p, since these binomial coefficients are integers, and the factors in their denominators are $<p$. Therefore

$$t_p \equiv 1 \pmod{p}. \tag{303}$$

Similarly

$$u_p = \frac{1}{2\sqrt{3}}[(1+\sqrt{3})^p - (1-\sqrt{3})^p]$$

$$= \frac{p}{1} + \frac{p(p-1)(p-2)}{1\cdot 2\cdot 3}\cdot 3 + \cdots + 3^{(p-1)/2}.$$

By Euler's Criterion, $3^{(p-1)/2} \equiv \left(\frac{3}{p}\right) \pmod{p}$. Therefore

$$u_p \equiv \left(\frac{3}{p}\right) \pmod{p}. \tag{304}$$

Now we use Eq. (300) with $n = p$, $m = 1$, and, since $t_1 = u_1 = 1$, we have

$$2u_{p-1} = t_p - u_p \equiv 1 - \left(\frac{3}{p}\right) \pmod{p}. \tag{305}$$

By Theorem 20, $\left(\frac{3}{p}\right) = 1$ if $p = 12m \pm 1$. Therefore for *these* primes we do get a "Fermat Theorem," since $p|2u_{p-1}$, and therefore $p|Q_{p-1}$. For the remaining primes $\neq 2$ or 3 we have $\left(\frac{3}{p}\right) = -1$. But from Eq. (298) we find

$$u_{p+1} = u_p + t_p \equiv 1 + \left(\frac{3}{p}\right) \pmod{p}. \tag{306}$$

Together with Eqs. (305), (304), and Theorem 79 we have thus proven

Theorem 80. *If p is an odd prime,*

$$p|Q_p, \qquad p|Q_{p-1}, \qquad or \qquad p|Q_{p+1}$$

according as $p = 3$, $12m \pm 1$, or $12m \pm 5$. Further if $p|Q_e$ and e is the smallest such positive index,

$$e = 3, \qquad e|p-1, \qquad or \qquad e|p+1$$

respectively.

Next we investigate the analogue of *Euler's Criterion*. From Eqs. (299) and (301)

$$6u_n^2 = t_{2n} - (-2)^n.$$

If $n = (p \pm 1)/2$ we use Eqs. (298) and (300) to obtain

$$-12u_{(p-1)/2}^2 = t_p - 3u_p + 2(-2)^{(p-1)/2},$$

$$6u_{(p+1)/2}^2 = t_p + 3u_p - (-2)^{(p+1)/2}.$$

Thus

$$-12u^2_{(p-1)/2} \equiv 1 - 3\left(\frac{3}{p}\right) + 2\left(\frac{-2}{p}\right) \pmod{p},$$
$$6u^2_{(p+1)/2} \equiv 1 + 3\left(\frac{3}{p}\right) + 2\left(\frac{-2}{p}\right) \pmod{p}. \tag{307}$$

We evaluate the Legendre Symbols—say from the table of $\left(\frac{-a}{p}\right)$ on page 47—and find:

If $p = 24m + 1$ or 11, $u^2_{(p-1)/2} \equiv 0 \pmod{p}$.

If $p = 24m + 13$ or 23, $3u^2_{(p-1)/2} \equiv 1 \pmod{p}$.

If $p = 24m + 5$ or 7, $3u^2_{(p+1)/2} \equiv -2 \pmod{p}$.

If $p = 24m + 17$ or 19, $u^2_{(p+1)/2} \equiv 0 \pmod{p}$.

In the first and last case $p|u_{(p\mp1)/2}$. In the two middle cases, since $p\nmid u_{(p\mp1)/2}$, while from Theorem 80, $p|u_{p\mp1}$, we see, from $u_{2n} = 2u_nt_n$, that $p|t_{(p\mp1)/2}$. We have therefore proven

Theorem 81. *Assume p prime. Then*

$$\begin{aligned} p|Q_{(p-1)/2} \quad &\textit{if} \quad p = 24m + 1, 11, \\ p|P_{(p-1)/2} \quad &\textit{if} \quad p = 24m + 13, 23, \\ p|P_{(p+1)/2} \quad &\textit{if} \quad p = 24m + 5, 7, \\ p|Q_{(p+1)/2} \quad &\textit{if} \quad p = 24m + 17, 19. \end{aligned} \tag{308}$$

These and similar results have been obtained by Lucas and by D. H. Lehmer.

EXERCISE 153.

$$P_p \equiv \left(\frac{2}{p}\right), \quad Q_p \equiv \left(\frac{6}{p}\right) \pmod{p}.$$

EXERCISE 154. For $n = p$ or $p \pm 1$, $P_n \equiv$ either ±1 or ±2, $Q_n \equiv$ either 0 or $\pm1 \pmod{p}$.

EXERCISE 155. Every prime Q_n except $Q_3 = 3$ ends in the digit 1.

63. THE LUCAS CRITERION

With the third case in Theorem 81 we have obtained that which we sought at the end of the last chapter. We analyzed Pepin's Theorem 55 there, and found that this test succeeded as a necessary and sufficient criterion for the primality of the Fermat number F_m because, in

$$F_m|(3^{(F_m-1)/2} + 1)(3^{(F_m-1)/2} - 1),$$

F_m divides only the first factor on the right, and also $F_m - 1$ is a power of 2. For M_p we have instead $M_p + 1$ as a power of 2. While Euler's Criterion is therefore useless our new "Euler Criterion" yields

Theorem 82 (Lucas Criterion). *A necessary and sufficient condition that $M_p > 3$ is prime is*

$$M_p|P_{(M_p+1)/2}. \tag{309}$$

This test may be carried out efficiently as follows. Let $S_1 = 4$, $S_2 = 14, \cdots,$ $S_{n+1} = S_n^2 - 2$. Then the condition becomes

$$M_p|S_{p-1}, \tag{310}$$

or, using residue arithmetic,

$$S_{p-1} \equiv 0 \qquad (mod\ M_p). \tag{310a}$$

EXAMPLES:

$$7 = M_3|P_4 = 7 \qquad 31 = M_5|P_{16} = 18817.$$

To test $M_7 = 127$ we use Eq. (310a) and arithmetic modulo 127. Then

$$S_1 = 4,\ S_2 = 14,\ S_3 \equiv 67,\ S_4 \equiv 42,\ S_5 \equiv 111,\ S_6 \equiv 0 \quad (\text{mod } 127).$$

For such a small M_p this test requires *more* arithmetic then Fermat's f_p and Euler's e_p on page 22. But consider M_{61}. Then e_{61} implies about a million divisions—and also a table of primes of the forms $488k + 1$ and $488k + 367$ out to 1.5 billion. However, Eq. (310a) requires only about 60 multiplications, 60 subtractions, and 60 divisions. Arithmetically speaking, a Lucas test for M_{61} is comparable with an Euler test for M_{31}, and a Cataldi test for M_{19}.

PROOF OF THEOREM 82. If $n = 2m + 1$, $M_n = 2^n - 1 \equiv 7 \pmod{24}$ by induction, since $M_3 = 7$ and

$$4(7 + 1) - 1 \equiv 7 \qquad (\text{mod } 24).$$

If M_p is prime for $p = 2m + 1$ we have Eqs. (309) by (308). Conversely, assume Eq. (309) and suppose a prime q divides M_p. Then $q|P_{(M_p+1)/2}$ and $q|t_{(M_p+1)/2}$. Since $u_{2s} = 2u_s t_s$ we obtain

$$q|Q_{M_p+1}.$$

Let e be the smallest positive integer where $q|Q_e$. By Theorem 79

$$e|M_p + 1 = 2^p.$$

If $e < 2^p$ we have $e|2^{p-1}$, and, by Theorem 78,

$$q|Q_{2^{p-1}} = Q_{(M_p+1)/2}.$$

This cannot be, since $q|P_{(M_p+1)/2}$, and $(P_s, Q_s) = 1$ for every s, since

$$P_s^2 - 3Q_s^2 = 1 \quad \text{or} \quad -2.$$

Therefore $e = 2^p$. But, by Theorem 80, the index e for any odd q satisfies $e \leqq q + 1$. Then $M_p = 2^p - 1 \leqq q$. Since $q|M_p$, we have $q = M_p$, that is, M_p is a prime.

Finally, since $P_2 + \sqrt{3}Q_2$ is the smallest solution of $x^2 - 3y^2 = 1$, we have

$$P_{2m} = P_m^2 + 3Q_m^2 = 2P_m^2 - 1$$

for any even m, by Eq. (292). If we define $S_n = 2P_{2^n}$ we therefore have $S_1 = 2P_2 = 4$, and $S_{n+1} = S_n^2 - 2$. Since $(M_p + 1)/2 = 2^{p-1}$, Eq. (310) is equivalent to Eq. (309).

We now give a brief account of the Mersenne numbers after Euler. There were then eight known Mersenne primes, the Greek primes:

$$M_2 = 3, \qquad M_3 = 7, \qquad M_5 = 31, \qquad M_7 = 127; \tag{311}$$

the medieval $M_{13} = 8191$; and the modern M_{17}, M_{19}, and M_{31}. Mersenne stated in 1644 that for $31 \leqq p \leqq 257$ there were only four such primes, M_{31}, M_{67}, M_{127}, and M_{257}. While Euler had verified M_{31} the remaining three were beyond his technique. There now ensued a pause of over a century.*

In 1876 E. A. Lucas used a test which is related to Theorem 82 and is described below. He found that M_{67} is composite and M_{127} is prime. With one or another of these Lucas-Lehmer criteria, and with extensive computations by hand or desk computers, all doubtful M_p were settled by the year 1947 for $31 < p \leqq 257$. It was found that

$$M_{61}, \quad M_{89}, \quad M_{107}, \quad \text{and} \quad M_{127}$$

are prime while the other M_p including M_{257} are composite.

The arithmetic necessary for a Lucas test of M_p is roughly proportional to p^3, since that in the multiplication of two n digit numbers is proportional

---

* Peter Barlow, in the article "Perfect Number" in *A New Mathematical and Philosophical Dictionary* (London, 1814), says "Euler ascertained that $2^{31} - 1 = 2147483647$ is a prime number; and this is the greatest at present known to be such, and consequently the last of the above perfect numbers, which depends upon this, is the greatest perfect number known at present, and probably the greatest that ever will be discovered; for as they are merely curious, without being useful, it is not likely that any person will attempt to find one beyond it."

to n^2. It is clear, then, that it becomes prohibitive to go much beyond $p = 257$ without a high-speed computer. The Lucas prime M_{127} therefore remained the largest known prime for three-quarters of a century. Further, a test of Catalan's conjecture was not possible. On the basis of Eq. (311), Euler's M_{31}, and Lucas's M_{127}, Catalan had "conjectured" that if $P = M_p$ is a prime then M_P is prime. If this were true, Conjecture 2 (and therefore Conjecture 1 also) would follow at once. But, for instance, is $M_{8191} = M_{M_{13}}$ a prime?

A. M. Turing in 1951 utilized the electronic computer at Manchester, England to test Mersenne numbers, but obtained no new primes. In 1952 Robinson used the SWAC in California and found five new primes:

$$M_{521}, \quad M_{607}, \quad M_{1279}, \quad M_{2203}, \quad M_{2281}.$$

There are no others for $127 < p < 2309$. In 1953 Wheeler used the ILLIAC and proved that M_{8191} is composite. The computation took 100 hours! Although it cannot be said that Catalan's conjecture was nipped in the bud, it was definitely nipped. It reminds one of the English philosopher Herbert Spencer, of whom it was said that his idea of a tragedy was "a theory killed by fact." In 1957 Riesel used the Swedish machine BESK to show that if $2300 < p < 3300$ there is only one more Mersenne prime, M_{3217}. Finally, in 1961, Hurwitz used an IBM 7090 to show that for $3300 < p < 5000$ there are two more Mersenne primes, M_{4253} and M_{4423}. The first of these is the first known prime to possess more than 1000 digits in its decimal expansion, while the twentieth known perfect number,

$$P_{20} = 2^{4422}(2^{4423} - 1),$$

is a substantial number of 2663 digits.

Exercise 156. The reduction of S_n^2 modulo M_p is facilitated by binary arithmetic. For let S_n modulo M_p be squared and equal $Q2^p + R$. If, therefore, R is the lower p bits of the square and Q is the upper p bits, then $S_n^2 \equiv Q + R \pmod{M_p}$. Or, if the right side here is $> M_p$, then $S_n^2 \equiv Q + R - M_p$. Thus the Lucas test requires *no* division if done in binary.

Exercise 157. (For those who know computer programming.) Estimate the computation time—say on an IBM 7090—to do a Lucas test on M_{8191}. (For those who have used desk computers.) Estimate the computation time—using residue arithmetic on a desk computer—to verify the following the following counter-example of Catalan's conjecture:

$$1 + 120\, M_{19} | M_{M_{19}}$$

that is,

$$2^{2^{19}} \equiv 2 \pmod{62914441}.$$

64. A Probability Argument

The Lucas test of M_{4423} on an IBM 7090 took about 50 minutes. It is clear that once again we are up against current limits of theory and technology. Suppose one had a computer 1,000 times as fast. Then one could test an M_p for p about 50000 in about one hour. However, there are about 10 times as many primes to be tested in each new decade, so that one would really want a computer 10,000 times as fast to do a systematic study out to $p \approx 50000$.

How many new Mersenne primes can be reasonably expected for $5000 < p < 50000$? A related question is this: Why do we call it Conjecture 2? Surely 20 Mersenne primes do not constitute "some serious evidence."

The answer is suggested by the prime number theorem:

$$\pi(N) \sim \int_2^N \frac{dn}{\log n}.$$

One can give a probability interpretation of this relation. However, it is not rigorous mathematics. The probability that an n chosen at random is prime is $1/\log n$. The heuristic argument goes as follows. Consider an interval of positive integers, $M - \frac{1}{2}\Delta M \leqq m \leqq M + \frac{1}{2}\Delta M$, with ΔM small compared with M, but large compared with $\log M$. Then the number of primes in this interval we estimate by

$$\int_{M-\frac{1}{2}\Delta M}^{M+\frac{1}{2}\Delta M} dm/\log m.$$

By the mean-value theorem this integral equals $\Delta M/\log(M + \epsilon)$ for a small ϵ. Thus the ratio of the number of primes to the number of integers here, which we call the probability, we may estimate as $1/\log M$.

Suppose now the Mersenne numbers M_p are *tentatively* considered numbers "chosen at random." Since $\log M_p \approx p \log 2$ the probable number of Mersenne primes M_p for $p_n \leqq p \leqq p_m$ would then be estimated by

$$P = \frac{1}{\log 2} \sum_{p_n}^{p_m} \frac{1}{p}.$$

The series on the right can be shown to be divergent, so that by choosing p_m large enough the probable number P could be made arbitrarily large. Now, in fact, the error in our assumption can only reinforce this conclusion. The "unrandomness" of the M_p is all in the direction of greater tendency towards primality. Thus $q \nmid M_p$ if $q < 2p + 1$. Again, any divisor of M_p is of the forms $2pk + 1$ and $8k \pm 1$, and all M_p are prime to each other. Everything we know suggests that our assumption errs on the conservative side.

Were such a "random" assumption valid it would follow, from the known rate of divergence of $\sum \frac{1}{p}$, that if M_{p_i} are the successive Mersenne primes, then $\log p_i$ would grow exponentially. Empirically, the sequence $p_i = 2, 3, 5, 7, 13, 17, 19, 31, 61, 89, 107, 127, 521, 607, 1279, 2203, 2281, 3217, 4253, 4423$ suggests a slower, linear growth of $\log p_i$. A reasonable guess is that there are about 5 new prime M_p for $5000 < p < 50000$.

We know much larger composite M_p than prime M_p. For example, as on page 29, $M_{16188302111}$ is composite. Primes of such a size are completely inaccessible to us with our current theory and technology. The Lucas test, when done in binary, appears so simple (see Exercise 156) that it may be hoped that one could penetrate more deeply into its meaning, and thereby effect the next breakthrough. Alternatively, however, it is also conceivable that one could obtain a (metamathematical ?) proof that the number of elementary arithmetic operations here is the minimum needed to decide the primality of M_p. But, to date, neither of these things has been done, and it is an Open Question which is the more likely.

Exercise 158. Give a heuristic argument in favor of infinitely many Wieferich Squares, $p^2 | 2^{p-1} - 1$. On the other hand, "explain" their rarity.

65. Fibonacci Numbers and the Original Lucas Test

Why do we single out $\sqrt{3}$ as a basis for a test; can we not use $\sqrt{5}$, say, instead? The answer is that the original Lucas test did use $\sqrt{5}$, via the so-called *Fibonacci numbers*.

Consider the continued fraction

$$x = 1 + \frac{1}{1+}\,\frac{1}{1+}\,\frac{1}{1+}\cdots$$

Since $x = 1 + (1/x)$, or $x^2 = x + 1$, we have

$$x = \tfrac{1}{2}(1 \pm \sqrt{5}),$$

but since $x > 0$ we must take the $+$ sign. The corresponding convergents to $(1 + \sqrt{5})/2$ are

$$\frac{1}{1}, \frac{2}{1}, \frac{3}{2}, \frac{5}{3}, \frac{8}{5}, \frac{13}{8}, \frac{21}{13}, \frac{34}{21}, \cdots.$$

The denominators (call them U_n) are the Fibonacci numbers. They are clearly definable by

$$U_1 = U_2 = 1, \qquad U_{n+1} = U_n + U_{n-1}. \tag{312}$$

The numerators are U_{n+1}, and we have

$$U_{n+1}/U_n \sim \tfrac{1}{2}(1 + \sqrt{5}). \tag{313}$$

It can be shown, by induction, that

$$U_n = \frac{1}{\sqrt{5}}\left[\left(\frac{1+\sqrt{5}}{2}\right)^n - \left(\frac{1-\sqrt{5}}{2}\right)^n\right]. \tag{314}$$

The analogue of Theorem 80 is

Theorem 83. *If p is an odd prime,*

$$p|U_p, \quad p|U_{p-1}, \quad \textit{or} \quad p|U_{p+1}$$

according as $p = 5$, $10m \pm 1$, or $10m \pm 3$. Further, if $p|U_e$, and e is the smallest such positive index,

$$e = 5, \quad e|p-1, \quad \textit{or} \quad e|p+1$$

respectively.

The original Lucas test was based on this Fermat-type theorem for $\sqrt{5}$. If $M = 10m - 3$, and $M|U_{M+1}$, but $M \nmid U_d$ for every divisor d of $M + 1$, it may be shown that M is a prime. Since

$$2^p - 1 \equiv -3 \pmod{10},$$

providing $p \equiv 3 \pmod 4$, the test is suitable for *one-half* of the Mersenne numbers, including M_{67} and M_{127}, but not M_{257}. By computing U_{2^p-1} and U_{2^p} modulo M_p one can determine the primality of the latter if $p \equiv 3$ (mod 4).

Lucas then modified this procedure into an Euler Criterion—type test as in Theorem 82. Let

$$V_1 = 1, \qquad V_2 = 3, \qquad V_{n+1} = V_n + V_{n-1}. \tag{315}$$

Then

$$V_n = \left(\frac{1+\sqrt{5}}{2}\right)^n + \left(\frac{1-\sqrt{5}}{2}\right)^n = U_{2n}/U_n.$$

If $R_n = V_{2^n}$, then it may be shown that

$$R_1 = 3, \qquad R_{n+1} = R_n^2 - 2.$$

It follows, if $p \equiv 3 \pmod 4$, that M_p is a prime if and only if

$$M_p|R_{p-1}.$$

Therefore Eq. (310) in Theorem 82 is also valid if we set $S_1 = 3$ instead of 4, but *only* if $p \equiv 3 \pmod 4$.

The difference between $\sqrt{3}$ and $\sqrt{5}$ as the basis of a test therefore comes to this—all M_p are of the form $12m + 7$, while some are of the form $10m + 1$, and others are of the form $10m + 7$. Another reflection of this difference is that all even perfect numbers, except the first, end in 4 when written in the base 12, but they end in 6 or 8 in decimal.

Exercise 159. Prove the results stated in this section. More generally, let $x_1 + \sqrt{N}y_1$ be the smallest solution of $x^2 - Ny^2 = 4$. Let $S_1 = x_1$ and examine the sequence $S_{n+1} = S_n^2 - 2$. Note that $x = z + (1/z)$ where $z = \frac{1}{2}(x + \sqrt{N}y)$. Specifically examine $N = 3$, 5, and 6, and develop a Lucas test based on $S_1 = 10$. Why can't $\sqrt{2}$ be used as the basis of a Lucas test? Relate this to the fact that the $\sqrt{2}$ exists in $\mathfrak{M}_{M_p}$—specifically, $(2^{(p+1)/2})^2 \equiv 2 \pmod{M_p}$.

Exercise 160. Use Eq. (232) with $N = 1$ and $a + bi = \cos\theta + i\sin\theta$ to derive the trigonometric addition laws for $\cos(\theta \pm \phi)$, etc. Interpret Eq. (244) as a generalized De Moivre's Theorem. Interpret the vectors (x_i, y_i) of Theorem 77 as an infinite cyclic group under the operation determined by Eq. (244). Reduce these vectors modulo a prime p and discuss the corresponding finite cyclic groups.

Exercise 161. (Lucas's Converse of Fermat's Theorem.) If $m | a^{m-1} - 1$, and $m \nmid a^d - 1$ for every divisor d of $m - 1$ which is $< m - 1$, then m is prime.

SUPPLEMENTARY COMMENTS, THEOREMS, AND EXERCISES

We utilize this section to tie down some loose ends developed in the foregoing three chapters, and also to give some further comments and exercises of interest in their own right. These results could have been included earlier, in the appropriate sections, but it seemed better not to attenuate the main argument. The 40 exercises which are given follow the order of the corresponding topics in the text.

EXERCISE 1S. On page 15 we noted a gap of 209 between successive primes. Show that there exist arbitrarily large gaps by considering the sequence $m! + k$ with $k = 2, 3, \dots, m$ for a large value of m (Lucas).

EXERCISE 2S. A less tricky, but also less simple proof of the foregoing result may be obtained by assuming the existence of a largest possible gap m, and showing that a consequence of this is contradicted by the Prime Number Theorem.

EXERCISE 3S. With reference to Conjecture 5, page 30, consider the sequence:

$$a_{i+1} = 2a_i + 1,$$

with $a_1 = 89$. Then a_1, $a_2 = 179$, $a_3 = 359$, $a_4 = 719$, $a_5 = 1439$, and $a_6 = 2879$ are all primes. But show that in any such sequence, regardless of the starting value a_1, the a_i cannot all be prime. In fact, infinitely many a_i must be composite.

EXERCISE 4S. ("Aus der ballistichen Zahlentheorie")

Two missiles, μ_1 and μ_2, are moving parallel to the x axis, and, at time $t = 0$, they pass each other in the following kinematic attitudes:

$$x_1(0) = x_2(0) = 0$$

$$\dot{x}_1(0) = \ddot{x}_2(0) = 1$$

$$\ddot{x}_1(0) = \dot{x}_2(0) = -1.$$

μ_1 has a sharp nose and many control surfaces, and therefore is decelerated by a skin-friction drag force which is proportional to its velocity. μ_2, a much older model (circa 1850), has a blunt nose and no control surfaces, and therefore is decelerated by an air-inertia drag force which is proportional to the square of its velocity.

$$V(t) = \frac{x_1(t) - x_2(t)}{t}$$

is the mean relative velocity, and is an analytic function of t. Show that the initial value of its n'th derivative, that is

$$\frac{d^n V}{dt^n}(0),$$

is an integer if and only if $n + 1$ is a prime.

EXERCISE 5S. Consider the seven sets of four residue classes b modulo 24 in the table on page 47. Omitting the residue 1 the remaining residues may be diagrammed as follows:

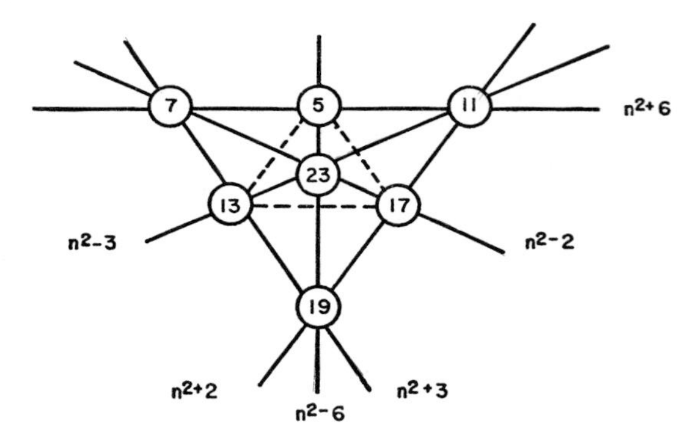

Six of the seven sets are shown as straight lines through three points. But the seventh set, $n^2 + 1$, is represented by the dotted line.

This would appear to give special roles to the form $n^2 + 1$ and the residue class in the center, $23 \equiv -1 \pmod{24}$. But *a priori* no residue class in $\mathfrak{M}_{24}$ except 1, and no subgroup of order 4, has a special role. Show, in fact, that any of the seven $n^2 + a$ may be given this "special role," and any b not on it may be placed in the center. There are thus 28 such diagrams.

But is there a configuration of seven straight lines and seven points, with each line on three points, and three lines through each point, so that we could draw a diagram with no $n^2 + a$ in a "special role"?

EXERCISE 6S. Show that Conjecture 12_1, on page 48, implies that

$$P_1(N) \sim P_4(N).$$

Compare with the empirical data.

EXERCISE 7S. Let N be written in decimal:

$$N = a_n 10^n + a_{n-1} 10^{n-1} + \dots + a_1 10 + a_0 .$$

Let the sum of the digits be

$$S_N = a_n + a_{n-1} + \dots + a_1 + a_0$$

and the alternating sum and difference be

$$D_N = a_n - a_{n-1} + \dots + (-1)^n a_0 .$$

Using residue algebra prove the divisibility criteria:

$$3|N \leftrightarrow 3|S_N ,$$

$$9|N \leftrightarrow 9|S_N ,$$

$$11|N \leftrightarrow 11|D_N .$$

EXERCISE 8S. (Gauss, Reciprocals, and Fermat's Theorem) On pages 53–54 we indicated that Gauss independently discovered Fermat's Theorem from his studies, as a boy, of a table of reciprocals. Let us *put ourselves in his place* and reconstruct his discovery. Gauss computed a table of reciprocals $1/m$ out to $m = 1000$. If $p(m)$ designates the period of $1/m$ in decimal, the period for all $m < 100$, and prime to 10, is given in the following table:

PERIOD OF $1/m$ IN DECIMAL

m	$p(m)$	m	$p(m)$	m	$p(m)$	m	$p(m)$
1	—	3	1	7	6	9	1
11	2	13	6	17	16	19	18
21	6	23	22	27	3	29	28
31	15	33	2	37	3	39	6
41	5	43	21	47	46	49	42
51	16	53	13	57	18	59	58
61	60	63	6	67	33	69	22
71	35	73	8	77	6	79	13
81	9	83	41	87	28	89	44
91	6	93	15	97	96	99	2

Now, being Gauss, the reader at once notes: $p(m) < m$ always; $p(m) = m - 1$ only if m is prime; but $p(m) \neq m - 1$ for every prime m; however, if $p(m) \neq m - 1$ for some prime m, then $p(m)|m - 1$ for that prime; and this implies $m|10^{m-1} - 1$ for every prime m other than 2 and 5.

Now prove this "conjecture" by noting, first, that the significance of the $p(m) = m - 1$, for some prime m, is, that during the division involving $p(m)$ digits, each remainder r (that is, residue of 10^n), from $r = 1$ to $r = m - 1$, occurs exactly once. If $p(m) = e < m - 1$ for some prime m,

e different remainders occur. If a is not one of these, a/m also has a period of e and *these* e remainders are all distinct from the foregoing. By continuation, and exhausting all possible remainders other than zero, $e|m - 1$. But the base 10 plays no essential role in the argument so that, for any prime m, $(a, m) = 1$ implies $m|a^{m-1} - 1$ or $a^{m-1} \equiv 1 \pmod{m}$.

Now, reader, relinquish your role as Gauss, resume that of a student and verify that Gauss's proof of Fermat's Theorem, in his book *Disquisitones Arithmeticae*, is essentially that which we have just reconstructed, and further, with a slight abstraction, this is the classic proof of Lagrange's Theorem (Exercise 71 on page 86) given in any book on group theory.

Note that whether one is led to Fermat's Theorem via the perfect numbers, or via periodic decimals, the problem does not *initially* concern itself with the concept of primality. The concept asserts itself, and enters the problem whether the investigator wishes it or not.

Exercise 9S. Let p be prime and $p \nmid a$. If $p \equiv 1 \pmod{4}$, a and $-a$ are both quadratic residues of p, or neither is. If $p \equiv -1 \pmod{4}$, exactly one of the two, a or $-a$, is a quadratic residue of p.

Exercise 10S. In Exercise 34, page 47, we saw that any value of $a = -1, \pm 2, \pm 3$, or ± 6 is a quadratic residue for one-half of the primes, and a quadratic nonresidue for one-half of the primes. Investigate this problem for all a. Start with any $a = (-1)^{(p-1)/2}p$, where p is an odd prime, such as $a = -3, +5, -7, -11, +13$, etc. Use Theorem 30 on page 63, and the Quadratic Reciprocity Law in the form preferred by Gauss:

$$\left(\frac{(-1)^{(p-1)/2}p}{q}\right) = \left(\frac{q}{p}\right), \tag{316}$$

and let q be a prime of the form $kp + b$. Now let $a = (-1)^{(M-1)/2}M$, where M is a product of distinct odd primes, and use Theorem 33 and its Corollary. Then let a be the negative of the foregoing, with q of the form $k\,(4M) + b$. Introduce a factor of 2 with $q = k\,(8M) + b$, and finally introduce any square factor.

Exercise 11S. Generalize the ideas in Theorem 33 to obtain the famous

Chinese Remainder Theorem. *Consider n moduli m_i prime to each other*:

$$(m_i, m_j) = 1 \qquad (i \neq j).$$

Then the set of congruential equations:

$$x \equiv c_i \pmod{m_i} \qquad (i = 1, 2, \ldots, n) \tag{317}$$

has a unique solution x modulo the product $M = m_1 \cdot m_2 \ldots m_n$. The solution may be obtained from the inverses:

$$a_i \equiv \left(\frac{M}{m_i}\right)^{-1} \pmod{m_i} \tag{318}$$

by the formula

$$x \equiv \sum_{i=1}^{n} a_i \frac{M}{m_i} c_i \ (mod\ M). \tag{319}$$

As an example find the four square roots of unity modulo 2047 = 23·89 by solving all four cases of

$$\begin{cases} x \equiv \pm 1 \pmod{23} \\ x \equiv \pm 1 \pmod{89}. \end{cases}$$

Further, two solutions of $x^2 - 2 \equiv 14 \pmod{2047}$ are obviously $x \equiv \pm 4$. Find two others.

EXERCISE 12S. Investigate the parallelism between the proofs of Theorems 34 and 36, both of which are due to Gauss. But also consider the significant difference whereby the $\phi(d)$ solutions x in the former theorem are given explicitly, while the $\phi(d)$ residue classes of order d in the latter are shown to exist nonconstructively.

EXERCISE 13S. If g is a primitive root of p, a prime of the form $4m + 1$, then so is $p - g$ a primitive root of p.

EXERCISE 14S. Show that the two proofs of the "if" part of Wilson's Theorem, that by Dirichlet, equation (52), and that of Exercise 54, page 74, are not as unrelated as they seem at first.

For the classical trick of summing $s = \sum_{n=1}^{p-1} n$ is to write the same sum backwards and *associate* integers with a common *sum*, thus:

$$\begin{array}{rl} s = & 1 + 2 + \cdots + (p-2) + (p-1) \\ s = & (p-1) + (p-2) + (p-3) + \cdots + 1 \\ \hline 2s = & (p-1) + (p-1) + (p-1) + \cdots + (p-1) + (p-1). \end{array}$$

On the other hand Dirichlet's proof associates integers with a common *product*, and one proof is a logarithmic version of the other.

As an aside the reader may note that the same "classical trick," abstractly speaking, is also at the foundation of Euclidean metric geometry. Euclid's I, 34 states that the diagonal of a parallelogram divides it into two equal parts:

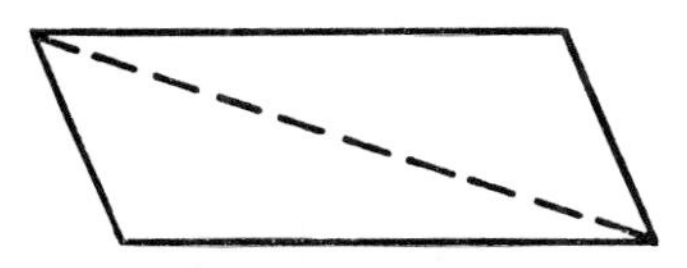

The parallel postulate comes in at I, 29, and the reader may verify, in the diagram on page 129, that all further consequences of I, 29 leading up to the Pythagorean Theorem utilize this I, 34.

EXERCISE 15S. A student, S. Ullom, notes in the diagram on page 75, that if we take differences modulo 17 we get the cyclic group again, rotated through a certain angle:

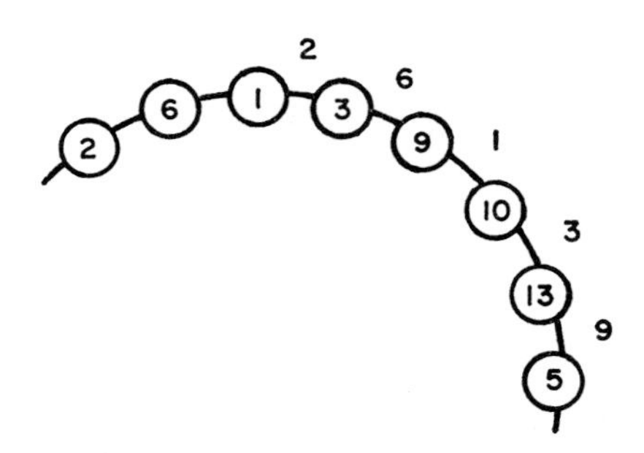

Prove that this property holds for every prime p and primitive root g.

EXERCISE 16S. In the definition of subgroup on page 83 it is redundant to stipulate that the set contains the identity. Further, if the group is *finite*, it is also redundant to stipulate the presence of every inverse. A subset of a finite group therefore is a subgroup of that group if it merely satisfies the closure postulate, (A) on page 60.

EXERCISE 17S. Ullom asks if the converse of Theorem 41 on page 85 is true. If all the squares in a group have an equal number of square roots is the group necessarily Abelian? Answer by A. Sinkov, no. There exists a non-Abelian group of order p^3 for every odd prime p wherein each element has one square root.

EXERCISE 18S. If $\mathfrak{M}_m \rightleftarrows \mathfrak{M}_{m'}$, their cycle graphs may be drawn so that that they look alike, i.e., they may be superimposed. Show that the converse is true; if they look alike, they are isomorphic.

EXERCISE 19S. To prove the criterion for the three-dimensionality of the cycle graphs of certain $\mathfrak{M}_m$ given on page 97 proceed as follows. First, note in $\mathfrak{M}_{63}$ the configuration involving the 3 square roots of unity other than 1, namely, 62, 8, and 55, and any three of the quadratic residues other than 1, say, 4, 25, and 37:

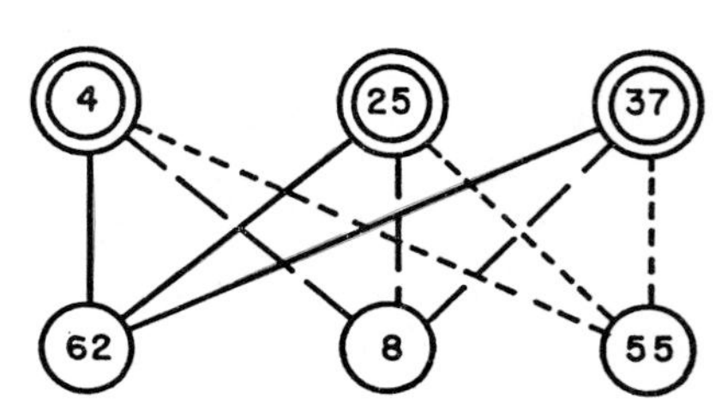

This portion (sub-graph) of the cycle graph is already three-dimensional (nonplanar). To see this, let us attempt to place these six residue classes in a plane and connect them without any crossing lines. First draw

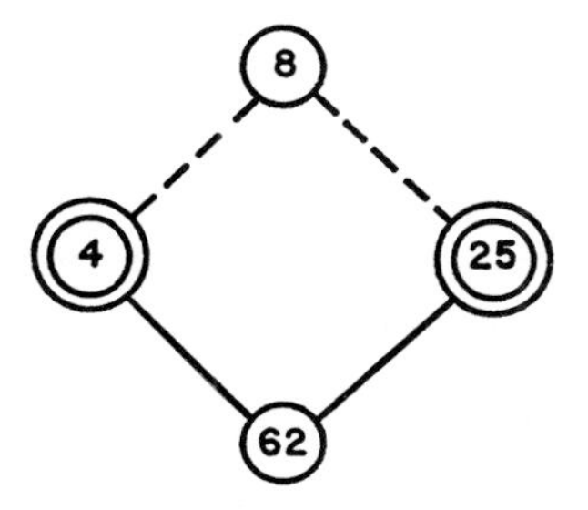

The path shown is a so-called *Jordan Curve*, and by the *Jordan Curve Theorem*, which see, the third quadratic residue can topologically go into only two places, the "inside" or the "outside." Any two points in the inside may be connected by a continuous arc lying wholly within the inside. Similarly for the outside. But if one point in the inside is joined to a point in the outside the connecting arc must cross the Jordan Curve. Choose the inside for the residue class 37 and connect to 8 and 62:

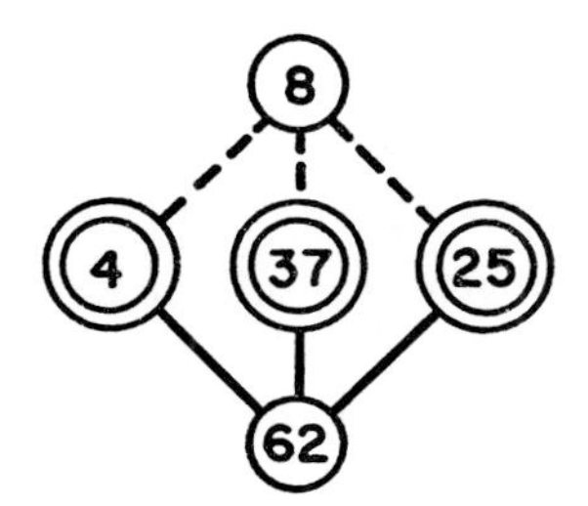

Now, by the Jordan Curve Theorem, we have three options for locating 55. Complete the proof that this sub-graph is nonplanar.

Since it is nonplanar it is clear that completion of the cycle graph, by adding other residue classes and lines, cannot undo this property, and therefore $\mathfrak{M}_{63}$ is also nonplanar.

Finally show that if $\mathfrak{M}_m$ has at least two characteristic factors, f_r and f_{r-1}, which are not powers of 2 the cycle graph of $\mathfrak{M}_m$ must contain a sub-graph similar to the foregoing and therefore $\mathfrak{M}_m$ is three-dimensional.

Exercise 20S. If, as on page 97,

$$\Phi_m = \langle 2^a \rangle \cdot \langle 2^b \rangle \cdots \langle 2^y \rangle \cdot \langle 2^z N \rangle$$

for some odd $N \geqq 1$, and likewise if

$$\Phi_{m'} = \langle 2^a \rangle \cdot \langle 2^b \rangle \cdots \langle 2^y \rangle \cdot \langle 2^z N' \rangle$$

with the same characteristic factors except for the last, prove that the cycle graphs of $\mathfrak{M}_m$ and $\mathfrak{M}_{m'}$ may be drawn so that the latter will contain N' lobes of the same structure as the N lobes in the former.

EXERCISE 21S. If $N = 1$ in the previous exercise we may say $\mathfrak{M}_m$ is *one-lobed*. Examples illustrated on pages 87–91 are $\mathfrak{M}_8$, $\mathfrak{M}_{15}$, $\mathfrak{M}_{24}$, $\mathfrak{M}_{64}$, $\mathfrak{M}_{85}$, and $\mathfrak{M}_{96}$. Gauss proved that an m-sided regular polygon may be constructed with a ruler and compass if, and only if, m is a power of two times a product of distinct Fermat primes. An m-sided regular polygon is therefore so constructable if, and only if, $\mathfrak{M}_m$ is one-lobed.

EXERCISE 22S. Prove the statement on page 98 that a cycle graph which contains four lobes of $\{2 \cdot 2\}$ does not represent a group since it implies a violation of the associative law.

EXERCISE 23S. On page 102 we indicated that the computations for obtaining the representation of $\mathfrak{M}_m$ from the primitive roots of the corresponding primes were indicated *explicitly* in the proof of Theorem 44, However, on page 99, only *one* of the two, h or $h + p$, was proven to be a primitive root of p^k, $k > 1$. Remove this *tentative* feature by showing, first, that if

$$h_n = h + np, \qquad (n = 0, 1, 2, \ldots p - 1)$$

the p values of h_n^{p-1} modulo p^2 are all incongruent and exactly one of them, say h_m, satisfies $h_m^{p-1} \equiv 1 \pmod{p^2}$. Now solve for m and give an explicit formula for h_{m+1}. The latter is a primitive root of p^k for all $k > 0$.

EXERCISE 24S. If the fallacious result in Exercise 80 on page 102 were true, it would follow from Exercise 79 that if 3 is primitive root of 487, it would not be a primitive root of 487^2. But show by computations like those on page 102 that 3 is a primitive root of both 487 and 487^2.

EXERCISE 25S. There are infinitely many primes of the form $12k - 1$.

EXERCISE 26S. The integer 2047 is a fermatian but not a Carmichael number. If $(a, 2047) = 1$, the probability that $a^{2046} \equiv 1 \pmod{2047}$ is $\frac{1}{4}$.

EXERCISE 27S. Prove Theorem 57 on page 138 without explicit reference to unique factorization. For if $(c, a) = g$, let $c = Cg$, $a = Ag$, and utilize $(A, C) = 1$.

EXERCISE 28S. Attempt to prove Theorem 63 on page 150 for the Gaussian integers. These are the algebraic integers in the field $k(e^{2\pi i/4})$—see page 152. If you succeed, try also $k(e^{2\pi i/3})$, and if you succeed here attempt to prove Conjecture 16 for the exponent 3. See page 152.

EXERCISE 29S. Attempt to prove Theorem 68 on page 162 by elementary means. Alternatively, investigate the elliptic theta functions and attempt to rederive Jacobi's proof mentioned on page 165.

EXERCISE 30S. Attempt to prove Theorem 75 on page 168.

EXERCISE 31S. (Euler's Identity) Unlike the result of Diophantus on page 159 for sums of *two* squares, $m = a^2 + b^2 + c^2$ and $n = e^2 + f^2 + g^2$ do *not* imply that $mn = j^2 + k^2 + l^2$. Find a counter-example. Derive from the vector algebra on page 169 the true relation:

$$\begin{aligned}(a^2 + b^2 + c^2)&(e^2 + f^2 + g^2) \\ &= (ae + bf + cg)^2 + (bg - cf)^2 + (ce - ag)^2 + (af - be)^2.\end{aligned} \tag{320}$$

But for *four* squares we again have a result analogous to that for two squares, for show that the last equation is a special case of *Euler's Identity*:

$$\begin{aligned}(a^2 + b^2 + c^2 + d^2)&(e^2 + f^2 + g^2 + h^2) \\ &= (ae + bf + cg + dh)^2 + (af - be + ch - dg)^2 \\ &\quad + (ag - bh - ce + df)^2 + (ah + bg - cf - de)^2.\end{aligned} \tag{321}$$

With reference to a textbook of modern algebra examine the parallelism between Diophantus's Identity and complex numbers on the one hand, and Euler's Identity and *quaternions* on the other.

EXERCISE 32S. (A false start on Theorem 61) In view of Euler's Identity in the previous exercise, if

$$p = w^2 + x^2 + y^2 + z^2$$

for every prime p, the Fermat-Lagrange Four-Square Theorem on page 143 would follow by induction. Now, by Theorems 60 and 72,

$$p = a^2 + b^2 + 0^2 + 0^2,$$

or

$$p = a^2 + b^2 + b^2 + 0^2$$

for all primes except those of the form $8k + 7$. Of the latter one-half are of the form

$$p = a^2 + b^2 + b^2 + b^2$$

by Exercise 129. Now attempt to express at least some of the remaining primes in the form

$$a^2 + b^2 + (2b)^2 + 0^2 \quad \text{or} \quad a^2 + b^2 + b^2 + (2b)^2$$

by use of Theorem 75. The attempt fails.

EXERCISE 33S. (Lagrange's Four-Square Theorem) A proof of Theorem 61 is known which is remarkably like that of Theorem 60 on page 159. There are small differences, due first, to the fact that Theorem 60 applies only to primes $\equiv 1 \pmod 4$ while Theorem 61 applies to all primes, and second, because $4(\frac{1}{2})^2 = 1$, while $2(\frac{1}{2})^2 < 1$. We first prove the

Lemma. *For every prime p there is a* q_0 *such that* $1 \leqq q_0 < p$ *and*

$$pq_0 = a_0^2 + b_0^2 + c_0^2 + d_0^2. \tag{322}$$

For $p = 2$ this is obvious, and for $p \equiv 1 \pmod 4$ we proceed as on page 159 with $c_0 = d_0 = 0$. For $p \equiv -1 \pmod 4$ let a be the smallest positive quadratic nonresidue of p. Then $a - 1$ and $p - a$ are both quadratic residues; and, by adding these, find an a_0 and b_0 such that

$$pq_0 = a_0^2 + b_0^2 + 1^2 + 0^2$$

with $q_0 < p$.

If q_0 is even in (322), 0, 2, or 4 of the integers there, a_0, b_0, c_0, and d_0, are even. By associating integers of the same parity, and renaming them if necessary, show that

$$pq_1 = p\left(\frac{q_0}{2}\right) = \left(\frac{a_0 + b_0}{2}\right)^2 + \left(\frac{a_0 - b_0}{2}\right)^2 + \left(\frac{c_0 + d_0}{2}\right)^2 + \left(\frac{c_0 - d_0}{2}\right)^2.$$

But if q_0 is odd proceed as in equation (217), page 159, etc., using the identity of Euler instead of that of Diophantus, and thereby obtain a q_1 less than q_0. Now complete the proof, again using (321).

The foregoing proof of the Lemma uses $(-1 \mid p) = (-1)^P$ for both classes of odd primes, $p = 2P + 1 \equiv \pm 1 \pmod 4$. A different proof uses the Box Principle on the $p + 1$ residue classes

$$x^2 \quad \text{and} \quad -1 - x^2$$

with $0 \leqq x \leqq P$. Then show that

$$x_i^2 \equiv -1 - x_j^2 \pmod p$$

for at least one i and j, and thus that

$$pq_0 = x_i^2 + x_j^2 + 1^2 + 0^2$$

for a $q_0 \leqq P$.

Exercise 34S. (Waring's Conjecture) The integer $7 = 4 + 1 + 1 + 1$ cannot be expressed as a sum of fewer than 4 squares. Similarly prove that

$$23 = 8 + 8 + 1 + 1 + 1 + 1 + 1 + 1 + 1$$

and

$$79 = 4 \cdot 16 + 15 \cdot 1$$

cannot be written as a sum of fewer than 9 positive cubes, and 19 biquadrates* respectively. Verify the generalization stated by one of Euler's sons that if

$$3^k = 2^k q_k + r_k \quad \text{with} \quad 0 < r_k < 2^k, \tag{323}$$

then the integer $2^k q_k - 1$ cannot be written as a sum of fewer than $I(k)$ positive k'th powers where

$$I(k) = 2^k + q_k - 2 = 2^k + \left[\left(\frac{3}{2}\right)^k\right] - 2. \tag{324}$$

Waring had earlier implied that *every* positive integer is the sum of $I(k)$ non-negative k'th powers.

A great deal of modern work in this direction has succeeded in "nearly" proving Waring's Conjecture. Hilbert proved that for every positive k there is a smallest $g(k)$ such that

$$n = \sum_{m=1}^{g(k)} x_m^k$$

for every positive n, with non-negative x_m. But he did not show that $g(k) = I(k)$, nor even give it an upper bound.

Wieferich proved that $g(3) = I(3) = 9$, and Pillai proved that $g(6) = I(6) = 73$. Dickson and Niven proved that if

$$r_k \leqq 2^k - q_k, \tag{325}$$

and $k \geqq 7$, then $g(k) = I(k)$. Verify (325) for $1 \leqq k \leqq 10$. It is now conjectured that (325) is true for all positive k, and, if this were true, Waring's Conjecture would be proven for every k except 4 and 5. If (325) is false for some k, and if $f_k = [(\frac{4}{3})^k]$, Dickson showed that, for that k,

$$g(k) = I(k) + f_k,$$

or

$$g(k) = I(k) + f_k - 1,$$

according as $2^k = f_k q_k + f_k + q_k$ or $2^k < f_k q_k + f_k + q_k$. Verify that $f_k q_k + f_k + q_k - 2^k > -1$ and therefore one of these two conditions must hold. In these cases, if any such exist, Waring's Conjecture would be false.

With reference to the ideas suggested by the cycle graph for $\mathfrak{M}_{64}$, show that, if $k \geqq 3$,

$$r_k \leqq 2^k - 5.$$

Estimate by heuristic probability considerations the probability that

* Biquadrate means fourth power.

(325) is violated for a particular k. Therefore show that the odds would favor the truth of Waring's Conjecture for all $k \geqq 10$.

There remain the hardest cases, $k = 5$ and $k = 4$. Dickson showed that

$$I(5) = 37 \leqq g(5) \leqq 54,$$

and Chandler showed that

$$I(4) = 19 \leqq g(4) \leqq 35.$$

It is curious that the easiest exponent for Fermat's Conjecture, namely, 4, is the hardest exponent for Waring's. Nonetheless the earliest and simplest result, due to Liouville, is for this exponent, and like the same exponent in Theorem 62 it utilized the theory for the exponent 2.

Liouville showed that $g(4) \leqq 53$, that is

$$n = \sum_{m=1}^{53} x_m^{\ 4}.$$

Let $n = 6q + r$, and, using the Four-Square Theorem,

$$n = 6a^2 + 6b^2 + 6c^2 + 6d^2 + r.$$

Again use this theorem on a, b, c, and d so that

$$n = 6\left(\sum_1^4 x_i^{\ 2}\right)^2 + 6\left(\sum_5^8 x_i^{\ 2}\right)^2 + 6\left(\sum_9^{12} x_i^{\ 2}\right)^2 + 6\left(\sum_{13}^{16} x_i^{\ 2}\right)^2 + r,$$

where $r = 0, 1, 2, 3, 4$, or 5. Now Liouville uses the identity:

$$\begin{aligned} 6\left(\sum_1^4 x_i^{\ 2}\right)^2 &= (x_1 + x_2)^4 + (x_1 - x_2)^4 + (x_1 + x_3)^4 + (x_1 - x_3)^4 \\ &+ (x_1 + x_4)^4 + (x_1 - x_4)^4 + (x_2 + x_3)^4 + (x_2 - x_3)^4 \\ &+ (x_2 + x_4)^4 + (x_2 - x_4)^4 + (x_3 + x_4)^4 + (x_3 - x_4)^4. \end{aligned} \tag{326}$$

Verify this identity and the proof follows at once.

When Liouville's recipe is applied to $n = 79$ we get exactly 19 positive biquadrates, not 53. We get, in fact, the representation on page 210.

The reader who wishes to pursue these problems will find a vast literature. There is not only the extensive analytic theory (due to Dickson et al) mentioned above, but Waring's Problem has also been extended to algebraic numbers by C. L. Siegel et al. There is also the so-called "easier" Waring's Problem. Note that if we allow negative integers, 23 is the sum of only 5 cubes:

$$23 = 3^3 + (-1)^3 + (-1)^3 + (-1)^3 + (-1)^3$$

More generally, a representation in the form

$$n = \sum_i \pm (a_i)^k$$

is allowed in this "easier" Problem (E. M. Wright et al).

Show that, with this new degree of freedom, the 4 squares necessary in Lagrange's Theorem may always be reduced to 3, e.g.:

$$28 = 14^2 - 13^2 + 1^2.$$

More generally, a representation as a sum and/or difference of three squares is also possible for those algebraic integers which may be written as a sum and/or difference of *any number* of squares (R. M. Stemmler). Examine the Gaussian integers and show that not all of them are representable as a sum and/or difference of squares.

EXERCISE 35S. (Theorem 76 for $N = -2$, see page 171)

Theorem 76$_1$ *Let $n \equiv \pm 1 \pmod 8$ and be > 1. If n is prime, $n = a^2 - 2b^2$ in a unique way in positive integers a and b such that $b \leqq \sqrt{n/2}$. Further $(a, b) = 1$. Conversely, if $n = a^2 - 2b^2$ in a unique way in non-negative integers with $b \leqq \sqrt{n/2}$, and if $(a, b) = 1$, then n is prime.*

First show, for any positive n, that if $n = a^2 - 2b^2$, and if $a > \sqrt{2n}$, we also have $n = a_1^2 - 2b_1^2$ with

$$a_1 = 3a - 4b,$$

$$b_1 = 3b - 2a,$$

and $0 < a_1 < a$. For the smallest $a > 0$ we therefore must have $a \leqq \sqrt{2n}$, and $b \leqq \sqrt{n/2}$ follows.

Show uniqueness, for n prime, somewhat as on page 160 with Eq. (232) instead of (215). From the analogue of Eqs. (222) and (223), and the inequalities for a and b, obtain a contradiction with a solution of

$$a^2 - 2b^2 = 1$$

having $0 < b < 2$.

Finally, one must prove the converse.

Further, for n prime, if x and y are obtained from Thue's Theorem, as in the proof of Theorem 74 on page 166, show that $2y - x = a$ and $y - x = b$ give the unique solution indicated in the above theorem. From

$$-M_p = 1 - 2(2^{(p-1)/2})^2$$

obtain a solution of

$$M_p = a^2 - 2b^2$$

with $b < \sqrt{M_p/2}$. Find such a representation for M_7 and *two* such for M_{11}.

Develop a result analogous to Theorem 76_1 for $a^2 - 3b^2$.

EXERCISE 36S. Using heuristic probability considerations similar to those used for Mersenne numbers on page 197 argue that there are only a finite number of Fermat primes as is suggested on page 80. Why is the argument less convincing in this case?

EXERCISE 37S. Obtain the constant (35a) for Conjecture 7 by a probability argument. If the probability of $n + 2$ being a prime were *independent* of the probability of n being a prime, we could assign $1/(\log n)^2$ as the probability that both are prime. But if $n > 2$, prime, and therefore if $2 \nmid n$, we automatically have $2 \nmid n + 2$. We therefore (tentatively) correct the probability to $2/(\log n)^2$ since on *this* ground, if n is known to be prime, $n + 2$ now has twice the probability. But, again, if $n > 3$, prime, and therefore $3 \nmid n$, $n + 2$ has 1 chance in 2 of being divisible by 3, not 1 chance in 3. We again correct to

$$2 \frac{1 - \frac{1}{2}}{1 - \frac{1}{3}} \frac{1}{(\log n)^2}.$$

By continuation, obtain (35a), and by integration obtain the conjectured asymptote in (35).

For large N it is known that the agreement in (35) is good. Thus D. H. Lehmer finds $Z(37 \cdot 10^6) = 183728$, while the right side of (35) for $N = 37 \cdot 10^6$ is $183582. \cdots$.

EXERCISE 38S. If $Z^{(k)}(N)$ is the number of pairs of primes of the form $n - k$ and $n + k$ for $n + k \leqq N$, advance an argument to show that

$$Z^{(2)}(N) \sim Z^{(1)}(N),$$

but

$$Z^{(3)}(N) \sim 2Z^{(1)}(N).$$

EXERCISE 39S. Develop a strong conjecture which bears the same relation to Conjecture 4 as Conjecture 7 does to Conjecture 6. Using the datum $Z(1000) = 35$ estimate the number of M_p, with $p < 1000$, for which $2p + 1 | M_p$. Compare with the list on page 28.

EXERCISE 40S. (Lucas Sequences) From page 199 the $S_1 = 4$ in Theorem 82 may be replaced by $S_1 = 3$ for one-half of the M_p. But in Exercise 159 it develops that $S_1 = 10$, like $S_1 = 4$, is valid for every M_p. Show that besides $S_1 = 4$, and $S_1 = 10$, there are infinitely many such *universal starters*. For instance 52 is one such, and if x is one, so is $x(x^2 - 3)$. Hint: Note, on page 188, that $4 = 2P_2$ while $52 = 2P_6$.

Study the transformation

$$x_{i+1} \equiv x_i^2 - 2 \qquad (327)$$

acting upon every residue class modulo a prime M_p. For M_5 verify the following diagrams:

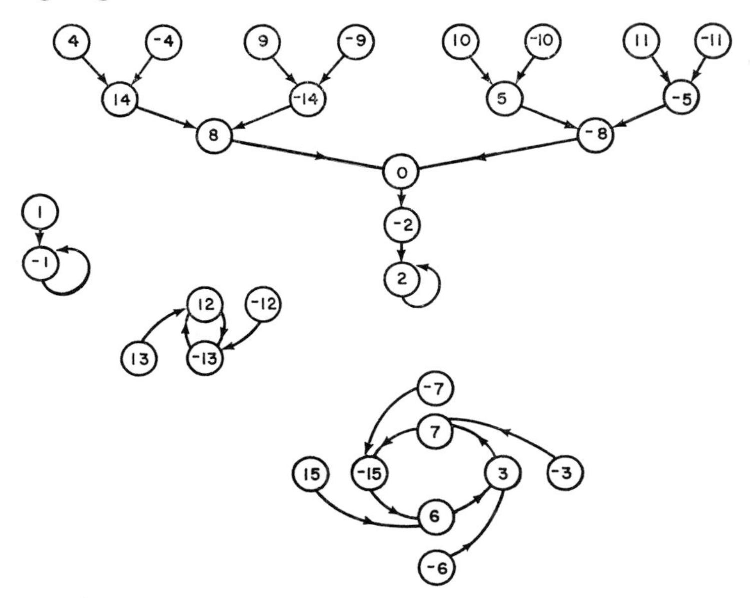

Here the $\rightarrow$ means application of the transformation (327).

Now note the following: $52 \equiv -10 \pmod{M_5}$. The repeated application of the transformation

$$x_{i+1} \equiv x_i(x_i^2 - 3) \qquad (328)$$

to any of the 8 possible starters in the top row of the main pattern gives a cyclic sequence of period 8 which runs through these 8 starters. Application of (328) to the second row gives the second row in a cycle of period 4, etc. Omitting the residues 0 and ± 2 all $\frac{1}{2}(M_5 - 3)$ of the remaining residues in the main pattern satisfy $\left(\dfrac{x^2 - 4}{M_5}\right) = -1$ while the $\frac{1}{2}(M_5 - 3)$ residues in the spiral patterns satisfy $\left(\dfrac{x^2 - 4}{M_5}\right) = +1$.

Develop a general theory for all prime M_p, proving the main theorems, if you can.

CHAPTER IV

PROGRESS

66. Chapter I Fifteen Years Later

First, read the Preface to the Second Edition. Square brackets below indicate references: [1]–[34] are the annotated references of the first edition, while [35]–[154] have been added for this chapter.

There has been work on Open Question 1, page 2. In [35] Hagis shows that no odd perfect number is less than 10^{50}. His long, detailed 83-page notebook [36] has been carefully checked by his principal competitor Tuckerman, and so we must accept it as valid. In [37] Buxton and Elmore claim 10^{200}, but I do not know that their proof has been similarly authenticated.

Does this 10^{50} bound change the status of Open Question 1 to that of a Conjecture? Not in my opinion; 10^{50} is a long way from infinity and all we can conclude is that there is no small odd perfect number. In fact, of the 24 known even perfect numbers, only the first nine are smaller than 10^{50}, so we cannot even state that $P_{10} = 2^{88}M_{89} \approx 1.9 \cdot 10^{53}$ is the tenth perfect number. When one examines the elaborate [36] it certainly seems doubtful that anyone will overtake $P_{24} = 2^{19936}M_{19937} \approx 9.3 \cdot 10^{12002}$ by such methods. But Hagis himself graciously implies [38] that Tuckerman's algorithm [39] may be more powerful than his.

There has been work on the table of $\pi(n)$, page 15. Lehmer's $\pi(10^{10})$ listed there is correct as shown, although [3] erroneously gave it as 1 larger. Bohman [40] worried about this discrepancy at length, but he then continued, using the same method, to compute

$$\pi(10^{11}) = 004118054813,$$

$$\pi(10^{12}) = 037607912018,$$

$$\pi(10^{13}) = 346065535898.$$

The gap of 209 consecutive composites on page 15 is the largest gap [4] that occurs up to 37 million. Skipping over intermediate work, which is referenced in Brent's [41], we find in [41] that the prime $p =$

2614941710599 is followed by 651 composites and that all gaps that occur before p are smaller. Every possible gap $1, 3, 5, \ldots$ up to 533 occurs below p, and its first occurrence has been recorded. The evidence in [41] and elsewhere supports the conjecture that I gave in [42] and I now wish to add

Conjecture 18. *Let $p(g)$ be the first prime that follows a gap of g or more consecutive composites. If all gaps that occur earlier are smaller than g we call g a maximal gap, and we have the asymptotic law*

$$\log p(g) \sim \sqrt{g} \tag{329}$$

as $g \to \infty$.

More general and stronger conjectures are discussed in [41] and in papers cited there.

Section 10 made the point, like it or not, that the perfect numbers had a great influence in the development of number theory. Aliquot sequences are closely related to perfect numbers. One iterates the operation

$$s(n) = \sigma(n) - n, \tag{330}$$

where $\sigma(n)$ equals the sum of the positive divisors of n. See [43] for an introduction. If $s(n) = n$, then n is perfect. Study of these sequences has surely been one of the causes of the many remarkable new developments in primality theory and in factorization methods that have occurred in recent years. So we see the same forces acting before our very eyes (at a lower level, to be sure). The reason is clear: the perfect numbers (always) and the aliquot sequences (frequently) grow very rapidly, and if one is to handle them one is constantly forced to invent stronger and stronger methods. The sequences $a^n \pm 1$ are also related, and a project for factoring them has been another cause of these new developments. Their exponential growth creates the same situation and, as before, Necessity becomes the Mother of Invention.

Now consider Conjecture 4 and Exercise 16 on page 29, and the answer to the latter on page 169. Exercise 39S calls for a stronger quantitative version of Conjecture 4, and we could also ask for a stronger modification of Exercise 16. The generalization was given in [44] and we call it

Conjecture 19. *Let $f_k(N)$ be the number of M_p with $p \leqq N$ that have a prime divisor $d = 2kp + 1$. Then*

$$f_k(N) \sim \bar{z}(N)\frac{\cos^2(k\pi/4)}{k}\prod_{q|k}\left(\frac{q-1}{q-2}\right)\left[1 - \frac{\log 2k}{\log N}\right] \tag{331}$$

as $N \to \infty$, *where* $\bar{z}(N)$ *is the right side of* (35) *and the product above is taken over all odd primes q, if any, that divide k.*

In [44] the conjecture is stated in a stronger form: the order of the error term is given. The heuristic arguments and data given in [44] make Conjecture 19 very plausible. We return to it presently.

Conjectures 6 and 7 about twin primes are truly key questions. The twin primes 140737488353700 ± 1 were the largest known to me in 1962 but one of the new primality criteria alluded to above has yielded [45] the much larger pair $76 \cdot 3^{139} \pm 1$. These primes have 69 decimal digits but no doubt even larger pairs could be found by the same method. Brent [46] (see also [47], [48]) has counted the twins up to 10^{11} and finds

$$z(10^{11}) = 224376048,$$

so that we could now give one pair to every American.

The evidence for Conjectures 6 and 7 is overwhelming, and although they remain unproved, interest has already shifted to the second-order term

$$r_3(N) = \bar{z}(N) - z(N). \tag{332}$$

This difference oscillates [46, Fig. 3] around zero in an unpredictable way; it is not understood at all [48].

In his famous paper [49] that initiated sieve theory, Brun proved that the series

$$B = \frac{1}{3} + \frac{1}{5} + \frac{1}{5} + \frac{1}{7} + \frac{1}{11} + \frac{1}{13} + \frac{1}{17} + \frac{1}{19} + \cdots \tag{333}$$

converges. The denominators here are the twin primes. The accurate computation of *Brun's constant* B is a real challenge [50]. Assuming (35), Brent [46] estimates

$$B = 1.9021604 \pm 5 \cdot 10^{-7}. \tag{334}$$

This is probably correct, or nearly correct, but the unpredictable $r_3(N)$ makes it very difficult to obtain greater accuracy. While B is a well-defined real number, its evaluation to, say, 20 decimals would not only require a proof of Conjecture 7 but would require the understanding of $r_3(N)$ besides.

For *all* primes, the analogous

$$r_1(N) = \int_2^N \frac{dn}{\log n} - \pi(N)$$

can be expressed in terms of the complex zeros of the Riemann zeta function [51]. That is bad enough, but for $r_3(N)$ we lack even that.

The generalization $Z^{(k)}(N)$ of Conjecture 7 referred to in Exercise 38S, which counts the prime-pairs

$$n - k \quad , \quad n + k \tag{335}$$

for $n + k \leqq N$, had been examined in [4] for $k = 1, 2, \ldots, 70$. The more difficult problem

$$p_{n+1} - p_n = 2k \tag{336}$$

concerning consecutive primes has been impressively studied by Brent. In [41] he estimated the value of p_{n+1}, where (336) is first satisfied, and in [52] he estimated the number of solutions of (336) for $p_{n+1} \leqq N$. His extensive empirical data convincingly agrees with the conjectures deduced there from reasonable heuristic arguments. Of course, none of these conjectures was proved.

Going beyond the linear polynomials (335) to Conjecture 12_1 and the table on page 49, let us add [53] as another source of data on $P_a(N)$ besides the earlier [16]. For $P_1(N)$ alone, that is, for primes of the form $n^2 + 1$, Wunderlich [54] has gone much further and we record his

$$P_1(10^6) = 54110 \quad \text{and} \quad P_1(10^7) = 456362.$$

As expected, they agree well with Conjecture 12.

The Bateman-Horn Conjecture [34] is a most important generalization. Briefly (but see [34]), if

$$f_1(n), f_2(n), \cdots f_k(n) \tag{337}$$

are k independent, irreducible polynomials in n, and if $Q(N)$ is the number of $n \leqq N$ for which all of the k $f_i(n)$ are simultaneously prime, then

$$Q(N) \sim C \int_2^N \frac{dn}{(\log n)^k} \tag{338}$$

as $N \to \infty$, where C depends upon the array (337) and is given by a very slowly convergent product.

The linear and quadratic cases above are all special cases of (337) and *all* other polynomials that have been studied, such as

$$f_1 = n^4 + 1, \quad f_1 = n^3 + 3, \quad f_1 = n^6 + 1091,$$

$$f_1 = (n-1)^4 + 1, \quad f_2 = (n+1)^4 + 1,$$

etc. have given results consistent with (338). An accurate computation of the appropriate C is frequently difficult, but in [55] Davenport and Schinzel give a useful first approximation. Recently [56], Epstein zeta functions have been found to be very effective in computing many such constants C accurately.

Except for the single linear polynomial $f_1 = an + b$, with $(a, b) = 1$, where (338) reduces to the (28) in de la Vallée Poussin's Theorem 16, no case of (338) has been proved. Nonetheless, one can be quite confident, for example, that although $f_1 = n^4 + 2$ has never been studied, one can now compute its C accurately (say, to 12 decimals) by Epstein zeta functions and would find that that C and $k = 1$ in (338) would accurately estimate its $Q(N)$ for large N.

An unusual result of Hensley and Richards may offer a different type of evidence. If the f_i in (337) are all linear, and if we assume infinitely many k-tuples of such primes for each suitable array (337), without requiring the stronger result (338), Hensley and Richards [57] show that for some integers x and $y \geqq 2$ we have

$$\pi(x + y) > \pi(x) + \pi(y). \tag{339}$$

Since this contradicts a frequently suggested property of $\pi(x)$, it would be desirable to find such a counter-example. There is none with $x = y$, since it was recently proved [57a] that $\pi(2x) < 2\pi(x)$ if $x \geqq 11$. While an example of (339) would certainly not prove the k-tuple prime conjecture, it would at least verify a predicted but unexpected consequence.

Goldbach's Conjecture 8 has been verified [58] by Stein and Stein for all even numbers up to 10^8. The historically important variant

$$4n + 2 = p_1 + p_2$$

with $p_1 \equiv p_2 \equiv 1 \pmod 4$, which was mentioned on page 244, was also verified to the same limit if we allow $p_1 = 1$ for a few small n. Hardy and Littlewood [9] also gave a strong version of Conjecture 8. If $P(2n)$ is the number of solutions of

$$2n = p_1 + p_2$$

then

$$P(2n) \sim \bar{z}(2n) \prod_{q|n} \left(\frac{q-1}{q-2}\right), \tag{340}$$

in the notation of (331). This has been satisfactorily verified to $n = 10^5$ in [59]. See also [60] for a different version.

The extensive development of sieve methods since Brun's time (cf. [61]) has been largely directed towards the proofs of weakened conjectures. The result that is closest to (340) was obtained by Jing-run Chen [62] (cf. [61]). He showed that, for all n greater than some n_0, the number of solutions of

$$2n = p_1 + P_2,$$

where P_2 is the product of at most *two* primes, is greater than one-third of the right side of (340).

As this is being written, there has just appeared [63] a result of Pogorzelski which states "The Goldbach Conjecture is provable from the following: The Consistency Hypothesis, The Extended Wittgenstein Thesis, and Church's Thesis." It seems unlikely that (most) number-theorists will accept this as a proof of Conjecture 8 but perhaps we should wait for the dust to settle before we attempt a final assessment.

Schinzel [64] has generalized the strong Goldbach conjecture (340). This complements the Bateman-Horn conjecture and, although it has not been studied as extensively as the latter, there is no reason to think that it is not equally reliable.

Finally, returning briefly to (331), we note that only for $k = 1$ is this contained in the Bateman-Horn Conjecture. For $k = 3, 4, 5, 7, \ldots$ (331) takes us into a new realm and thus suggests [44] that the Bateman-Horn conjecture can and should be generalized further. Also of interest in [44] is the speculation there that it may now be possible to prove Conjecture 3, which states that there are infinitely many Mersenne composites. Of all the conjectures in Section 12, Conjecture 3 is certainly the hardest to doubt and perhaps the easiest to prove. It is embarrassing that none of the conjectures in Section 12 are yet proved and good strategy therefore suggests a serious attack on Conjecture 3.

Exercise 162 ("Hard Times"). In the 4000 numbers $f_1(n) = n^6 + 1091$, ($n = 1$ to 4000), there is only one prime. Identify it, and estimate the small constant C in (338) for this $f_1(n)$; [65], [66].

67. Artin's Conjectures, II

Artin's Conjecture 13 remains as before: there is little doubt that it is true, but it has not been proved, not even for a single base a, including the values $a = -2, -4, +3$ cited in Theorems 38–40. For $a = 3$ in Theorem 40, while many new factors have been found for various F_m, [67], [68], [69] including the spectacular factorization

$$F_7 = 59649589127497217 \cdot 5704689200685129054721,$$

not a single new prime F_m has been found and perhaps there are none. Nonetheless, there is little doubt that Conjecture 13, and Conjecture 14 also, are true for $a = 3$.

But Conjecture 14 is not true, as it was stated, for *all* $a \neq b^n$ with $n > 1$; in particular, it is false for $a = 5$. The heuristic argument for $a = 2$ on page 82 is sound, and it also applies to $a = 3$. But for $a = 5$ it is not sound; Artin has an oversight here and we have followed him too uncritically. Those p that have 5 as a quintic residue, i.e., those for which one has $5|G$ in the notation above, were deleted there by multiplying by the factor

$$\left(1 - \frac{1}{5 \cdot 4}\right).$$

But these p are all $\equiv 1 \pmod 5$ and since $(5|5k + 1) = +1$ by the Reciprocity Law, they also have $2\,|G$ and we have *already deleted them*, with the factor

$$\left(1 - \frac{1}{2}\right).$$

For $a = 5$, being a quintic residue is not independent of being a quadratic residue. That is the only erroneous factor for $a = 5$, and so we should expect

$$\nu_5(N) \sim \frac{20}{19} A\pi(N) \tag{341}$$

instead of (117). Therefore, $a = 5$ should have a density of primitive roots that is about 5% higher.

What is really embarrassing here is that it is just what one finds in Cunningham's table on page 81! We accepted the high $\nu_5(10{,}000) = 492$ there because it exceeded $A\pi(10{,}000) = 459.6$ by less than $2\sqrt{459.6}$ and by an imprecise probability estimate such an excess seems to be an acceptable fluctuation. If Cunningham had continued his table for $a = 5$ until $N = 10^5$ or 10^6 the error would certainly have become obvious.

For the seven other a in Cunningham's table, Conjecture 14 needs no change. But for $a = 13, 17, 29, \ldots$ or $a = -3, -7, -11, \ldots$, that is, for any prime $\equiv 1 \pmod 4$, we have the same coupling between $2|G$ and $|a|\,|G$, and (341) generalizes to

$$\nu_a(N) \sim \frac{|a| \cdot (|a| - 1)}{|a|(|a| - 1) - 1} A\pi(N). \tag{342}$$

Had Cunningham computed the data for $a = -3$, the fact that its density runs 20% higher than that for $a = 2$ and 3 would surely have exposed the error much earlier.

D. H. and Emma Lehmer discovered and analyzed these errors in their aptly entitled paper "Heuristics, Anyone?" [70], where they did include data for $a = -3$. For most small a the correction needed for (117), if any, is rather obvious; but the general case is somewhat complicated, and for brevity we refer the reader to Heilbronn's formulation in [71, secs. 23, 24].

EXERCISE 163. Show that for $a = -15$ there is coupling between the cubic and quintic residues and therefore the conjecture should be $\nu_{-15}(N) \sim 94A\pi(N)/95$.

Let us now record

Conjecture 14 (Amended). *If a is not -1 or a square, then*

$$\nu_a(N) \sim f_a A\pi(N), \tag{343}$$

where f_a is a rational number given by Heilbronn's rules [71]. *Frequently, e.g., for $a = -6, -5, -4, -2, 2, 3, 6$, f_a is simply equal to 1.*

The next big development was that Hooley [72] proved Conjecture 14 (Amended) *conditionally*. He showed that (343) follows if one assumes that the Riemann Hypothesis holds for certain Dedekind zeta functions. (Clearly, that implies that Conjecture 13 also follows under these conditions.) His proof goes well beyond our subject matter and we confine ourselves to one remark: Hooley's bound for the error term

$$|\nu_a(N) - f_a A\pi(N)|$$

is rather large compared with the known empirical data.

Baillie computed both sides of (343) for all a between -13 and $+13$ inclusive and all N up to $33 \cdot 10^6$. In my review [73] of this extensive table, I point out that

$$|\nu_a(N) - f_a A\pi(N)| < \sqrt{f_a A\pi(N)} \tag{344}$$

is valid for all a and N in this range. While we certainly do not know that (344) remains valid for larger N, this does seem to suggest that it may be possible to reduce Hooley's error term, assuming, as before, all needed Riemann Hypotheses.

An elementary variation [74] on (343) of interest is given by

Definition 42. If g is a primitive root of p that satisfies

$$g^2 \equiv 1 + g \pmod{p}, \tag{345}$$

we call g a *Fibonacci primitive root.*

Since (345) implies

$$g^3 \equiv g + g^2, \quad g^4 \equiv g^2 + g^3, \quad \text{etc.} \pmod{p}, \tag{346}$$

the sequence $g^0 \equiv 1, g^1 \equiv g, g^2, \ldots,$ which would normally be computed by repeated multiplication by $g \pmod{p}$, can also be computed *additively* by (346). An example is $g = 8$ for $p = 11$, and we have $g^2 \equiv 1 + 8 = 9$, $g^3 \equiv 8 + 9 \equiv 6$, etc. Now we state

Conjecture 20. *If $\nu_F(N)$ is the number of primes $\leqq N$ that have a Fibonacci primitive root, then*

$$\nu_F(N) \sim \frac{27}{38} A\pi(N) \tag{347}$$

as $N \to \infty$.

It was suggested in [74] that Hooley's conditional proof of (343) could probably be modified to be a conditional proof of (347), and this was recently done by Lenstra [75].

68. Cycle Graphs and Related Topics

On page 84 we indicated that $\mathfrak{M}_n$ and $\mathfrak{M}_{n+1}$ are isomorphic for $n = 3$, 15, and 104. This sequence continues with $n = 495, 975, \ldots$. For $n < 10^8$ there are twenty-three examples, the last of which is $n = 48615735$ (verify). It is not known whether the sequence is infinite, and that is also true of the much larger set of n for which $\phi(n) = \phi(n + 1)$. The latter condition is necessary but not sufficient; for $n < 10^8$ there are 306 examples [76].

The cycle graphs have proved to be useful when working with finite Abelian groups; and I have used them frequently in finding my way around an intricate structure [77, p. 852], in obtaining a wanted multiplicative relation [78, p. 426], or in isolating some wanted subgroup [79]. Any two Abelian groups that have superimposable cycle graphs are isomorphic, as in Exercise 18S. That is true for any groups, Abelian or not, that are of order < 16; but for order 16 one can display an Abelian and a non-Abelian group that have the same (abstract) cycle graph [80]. The non-Abelian one gives a nicer example for Exercise 17S, since its two square elements each have eight square roots. There is a second pair of such nonisomorphic look-alike groups among the fourteen groups of order 16.

Cyclic groups have such a simple structure that one is surprised when they yield an important new application. In many problems, one wants and needs a very efficient solution of

$$x^2 \equiv a \pmod{p}. \tag{348}$$

If $p = 4m + 3$, the answer is $\sqrt{a} \equiv a^{m+1}$, as in Exercise 47. But suppose $p = 8m + 5$ or (harder) $p = 8m + 1$. The importance of (348) was obvious to Gauss [81, p. 373] and to his best English expositor Mathews [82, p. 53] but neither came up with a particularly efficient method. Sometimes an efficient method is absolutely essential. In [77, p. 847] I am analyzing a certain subgroup and must solve (348) for $p = (2^{61} + 3)^2 - 8$, a prime of 37 digits. Unless the algorithm is highly efficient, that is impossible. But when one analyzes the location of a in the cyclic group $\mathfrak{M}_p$, a very efficient algorithm is not difficult to construct. For brevity, the reader is referred to [83, sec. 5].

Gauss's book finally got translated into English [84] but unfortunately the translation was not the best possible [84a]. The German edition, which contains considerable additional material, has been reprinted [81].

This year (1977) Gauss is 200 years old and I am much tempted to have a longish section discussing him, his work, and even his errors. But we have more pressing topics and for brevity we'll move on.

69. Pseudoprimes and Primality

What we called a fermatian in Definition 32 is usually called a 2-*pseudoprime* in the literature. Let us write

Definition 43. If

$$a^{n-1} \equiv 1 \pmod{n}, \tag{349}$$

n is called an *a-pseudoprime* whether it is composite or not. We abbreviate this as *a-psp*. Let $C_a(N)$ be the number of composite a-psp not exceeding N. If

$$a^{(n-1)/2} \equiv (a|n) \pmod{n}, \tag{350}$$

where $(a|n)$ is computed as if n were prime, n is called an *Euler a-psp*; we let $E_a(N)$ be the number of these that are composite. Let $c(N)$ be the number of Carmichael numbers.

Poulet's [23] dates from the pre-computer age and has many errors. Our table on page 117 reflects all the corrections known at the time of our first edition, but further errors have been found subsequently [85], [86]. Sam Wagstaff has now gone much further, and Poulet's table should be retired. We show an excerpt from Wagstaff's data [87]. I have included the ratio $C_2(N)/\sqrt{\pi(N)}$ from our inequality (156), $E_2(N)$ as far as I computed it on an HP-65, and $S_2(N)$ (which is defined later).

N	$C_2(N)$	$C_2(N)/\sqrt{\pi(N)}$	$E_2(N)$	$S_2(N)$	$c(N)$
10^3	3	0.231	1	0	1
10^4	22	0.628	12	5	7
10^5	78	0.796	35	16	16
10^6	245	0.874	112	46	43
10^7	750	0.920	—	162	105
10^8	2057	0.857	—	488	255
10^9	5597	0.785	—	1282	646
10^{10}	14885	0.698	—	3291	1547

Note that (156) remains valid in this much extended range; $C_2(N)/\sqrt{\pi(N)}$ has maxima near $N = 3 \cdot 10^6$ and $11 \cdot 10^6$ that are < 1, and it then falls steadily. The earlier (156) suggested Conjecture 15, but that conclusion had already been proved by Erdös [88]. We have

Theorem 84 (formerly Conjecture 15). *Almost all 2-pseudoprimes are prime.*

Erdös proved that

$$C_2(N)e^{(\log N)^{1/4}/3}/N \quad \text{is bounded.} \tag{351}$$

Therefore

$$(C_2(N) \log N)/N$$

must approach 0, and the theorem is proved.

But (351) clearly does not prove the much stronger (156) and, in fact, Erdös has repeatedly conjectured (cf. [89]) that $C_2(N)/N^{1-\epsilon}$ and even $c(N)/N^{1-\epsilon}$ will increase without bound for every positive ϵ. If he is correct, $C_2(N)/\sqrt{\pi(N)}$ will stop decreasing at some N and then will increase without bound. What is that N?

The matter is of interest. If a 40-digit n is a 2-psp, and if (156) holds, the probability that n is composite is less than 10^{-19}. But if $C_2(N)/\sqrt{\pi(N)}$ increases without bound starting at some unknown N, we lose that estimate. Erdös's "conjecture" remains controversial; it is not a conjecture as we defined it on page 2.

John Selfridge [87] has improved the subject with his

Definition 44. If $n = t \cdot 2^s + 1$ with t odd, n is a *strong a-psp* if

$$a^t \equiv \pm 1 \pmod{n} \qquad \text{or}$$

$$a^{t2^r} \equiv -1 \pmod{n}$$

for some positive $r < s$. Let $S_a(N)$ be the number of composite strong a-psp that do not exceed N.

Note that when one computes $a^{n-1} \pmod{n}$ one first computes $a^t \pmod{n}$ and then squares this residue s times. Any x that we thus encounter which satisfies $x^2 \equiv 1$ must equal ± 1 if n is a strong a-psp just as it does if n is a prime.

Exercise 164 (Selfridge). If n is a strong a-psp it is also an Euler a-psp. The two concepts are equivalent if $n \equiv 3 \pmod{4}$ but not if $n \equiv 1 \pmod{4}$.

Selfridge and Wagstaff have found that

$$N_1 = 2047 = 23 \cdot 89$$

is the first composite strong 2-psp, that

$$N_2 = 1373653 = 829 \cdot 1657$$

is the first composite strong a-psp for both $a = 2$ and 3, that

$$N_3 = 25326001 = 2251 \cdot 11251$$

is the first for $a = 2$, 3 and 5, and that

$$N_4 = 3215031751 = 151 \cdot 751 \cdot 28351$$

is the *only* composite strong a-psp for $a = 2$, 3, 5 and 7 that does not exceed $25 \cdot 10^9$.

EXERCISE 165. Show that N_4 is a Carmichael number. Show that N_4 is a strong a-psp for $a = 2$, 3, 5, and 7 but not for $a = 11$ simply by showing that

$$\left(\frac{a}{151}\right) = \left(\frac{a}{751}\right) = \left(\frac{a}{28351}\right)$$

is true for $a = 2$, 3, 5 and 7, but not for 11.

EXERCISE 166. Examine the cycle graph of the subgroup $C_4 \times C_8$ in $\mathfrak{M}_{N_2}$. If $(a, N_2) = 1$, the probability that N_2 is an a-psp is 16/32; the probability that it is an Euler a-psp is 8/32 and the probability that it is a strong a-psp is 6/32. N_2 is an Euler 67-psp but not a strong 67-psp.

Our table of $C_2(N)$, etc. suggests several questions, all of which are open. We note that $E_2(N)/C_2(N)$ is running a little less than 1/2, but we do not know what happens as $N \to \infty$. (We should emphasize that this ratio is an average: for $n \equiv 1 \pmod 8$ alone the fraction is much larger.) It is probable, but unproved, that $c(N)/C_a(N) \to 0$. It is plausible, but unproved, that $S_a(N)/C_a(N) \to 0$ very slowly, say as $(\log \log N)^{-1}$.

In contrast with Erdös's $C_2(N)/N^{1-\epsilon}$, even $C_2(N)/\log N$ has not been shown to increase without bound. Nonetheless, we list

Conjecture 21. *The ratio $C_a(N)/N^{1/2-\epsilon}$ increases without bound for all a and any positive ϵ.*

For consider the numbers

$$n(m) = (12m + 1)(24m + 1), \tag{352}$$

where both factors on the right are prime. Then $n(3)$ is the 10th composite 2-psp on page 117 and $n(69)$ gives Selfridge's N_2 above. Since $(2|24m + 1) = (3|24m + 1) = 1$, Theorems 44 and 46 show that each $n(m)$ is a 2-psp and a 3-psp. How many such $n(m)$ are there $< N$?

EXERCISE 167. Adapt the heuristic argument in Exercise 37S to these $n(m)$. Then the desired number should be asymptotic to

$$1.3203\sqrt{2N} \,/\, (\log N)^2, \tag{353}$$

where the coefficient is that in (35a). Show that the 25th number in (352) is $n(213)$ and $N = n(213)$ in (353) gives 25.14. Show that the 50th number is $n(519)$, and now (353) gives 49.84. Not bad.

Additional 3-psp are generated by

$$n'(m) = (12m + 7)(24m + 13),$$

and clearly these are not 2-psp.

For every a,

$$n_a(m) = (6am + 1)(12am + 1)$$

is both an a-psp and a 3-psp, so that there is little doubt that Conjecture 21 is true.

If (156) remains true (or nearly true) as $N \to \infty$, (353) shows that $C_2(N)$ is neatly trapped between $\sqrt{N}/\log^2 N$ and $\sqrt{N}/\sqrt{\log N}$. However, there is insufficient evidence to designate (156) a conjecture, and we are aware of Erdös's opinion. Numbers at infinity are quite different from those that we see down here: the average number of their prime divisors increases as $\log \log N$ and, while that increases very slowly, it increases without bound. People say that Erdös understands these numbers. We do note that the Erdös construction [89] that is said to yield so many Carmichael numbers is decidely peculiar in that they all are products of primes r_i for which each $r_i - 1$ is square-free. That is most untypical of the known Carmichael numbers; among the first 300 only three have that character, namely:

$$67 \cdot 331 \cdot 463, \qquad 23 \cdot 43 \cdot 131 \cdot 859, \qquad 131 \cdot 571 \cdot 1871.$$

All told, we regard the Erdös conjecture as an (unlisted) Open Question.

The n in (352) are *not* Carmichael numbers, since n is not an a-psp for any a that satisfies $(a|24m + 1) = -1$. The numbers

$$n(m) = (6m + 1)(12m + 1)(18m + 1)$$

are all Carmichael numbers if the three factors are prime, since

$$n(m) - 1 = 36m(36m^2 + 11m + 1).$$

Therefore, [90, p. 199] although it remains unproved that there are infinitely many Carmichael numbers, there is little doubt that $c(N)$ increases at least as fast as $CN^{1/3}/(\log N)^3$ for some constant C.

The Wieferich Squares (page 116) are much rarer; for $p < 3 \cdot 10^9$ there are still only the old examples of Meissner and Beeger [91].

As we indicated above, primality and factorization theory have advanced greatly in recent years. An exposition would require a whole book, and we merely give some key references here. If n is a strong a-psp for $a = 2$, 3, 5 and 7, then n is a prime if it is $< 25 \cdot 10^9$ and $\neq N_4$. But this is based on Wagstaff's table, which required much computer time and is therefore not extendable to very large n.

As an example, consider c_{937} in Theorem 58. It arises in the analysis of a certain simple group [92] and it is essential there that it be prime. But

c_{937} has 359 decimal digits and it surely would have defied all techniques known prior to the recent developments. A sketch of its primality proof is in [92, sec. 4]. The key reference is [93], an important paper of Brillhart, Lehmer and Selfridge. To be very brief, this combines generalizations of our Exercise 161 on page 200 and of our Theorem 82. It uses known factors of both $n - 1$ and $n + 1$, together with a bound B such that $n \pm 1$ have no other prime divisors $< B$, and combines all this into a powerful primality criterion for n. This has been implemented in computer programs and it is now routine to prove primality for large primes of, say, 50 digits. Our c_{937} is much larger, but its algebraic source (172) greatly assists us in factoring $c_{937} \pm 1$, and that suffices.

Besides the references in [93], which includes Pocklington, Robinson, Morrison, Riesel, etc., other pertinent references are Williams [94], [95], [95a] and Gary Miller [96]. The last contains an idea related to strong pseudoprimes. Certain factorization methods that give a *complete* factorization may also be used for primality tests if n is not too large. We return to them later; see [65], [78], [97].

In contrast to these highly technical, but very effective, methods we close this section with a new necessary and sufficient condition for primality that has more charm than utility [98].

Consider Pascal's Arithmetical Triangle with each row displaced two places to right from the previous row. The $n + 1$ binomial coefficients of $(A + B)^n$ are $\binom{n}{k}$, $k = 0, 1, \ldots, n$, and are found in the n-th row between columns $2n$ and $3n$ inclusive. Each coefficient in the n-th row is printed in bold-face if it is divisible by n.

Then we have

Column No.

Row No.	0	1	**2**	**3**	4	**5**	6	**7**	8	9	10	**11**	12	**13**	14	15	16	**17**	18	**19**	20	21	22	**23**
0	1																							
1			**1**	**1**																				
2					1	**2**	1																	
3							1	**3**	**3**	1														
4									1	4	6	**4**	1											
5											**1**	**5**	**10**	**10**	**5**	1								
6													**1**	**6**	15	20	15	**6**	1					
7															1	7	**21**	**35**	**35**	**21**	7	1		
8																	1	**8**	28	**56**	70	**56**	28	8
9																			1	**9**	**36**	84	**126**	**126**
10																					1	**10**	45	**120**
11																							1	**11**

Row No.

Theorem 85. *The column number is a prime if and only if all coefficients in it are printed in bold-face.*

For a proof, see [98].

70. Fermat's Last "Theorem," II

The ratios

$$\frac{144}{367} = 0.392 \quad \text{and} \quad \frac{72}{183} = 0.393$$

on page 153 are very suggestive; they are nearly equal and one asks: What is this number? Since a prime p must pass a gauntlet of $(p - 3)/2$ numbers B_n in Definition 40 (page 153) in order to be regular, we may heuristically estimate the probability P of regularity by

$$P = \left(1 - \frac{1}{p}\right)^{(p-3)/2} \tag{354}$$

if we assume that the numerators of the B_n are equidistributed (mod p). Then $P \sim e^{-1/2} = 0.60653$ as $p \to \infty$, and the density of irregular primes is therefore given by

Conjecture 22 (Lehmer [99], Siegel [100]). *If $I(N)$ is the number of irregular primes $\leqq N$ then*

$$I(N) \sim (1 - e^{-1/2})\pi(N) = 0.39347\pi(N) \tag{355}$$

as $N \to \infty$.

If Conjecture 22 were true, then by Theorem 64, Conjecture 16 would be true for at least three-fifths of all prime exponents.

Conjecture 16 itself is now true for all exponents $\leqq 125000$ by Wagstaff's calculations [101]. Further, he gives $I(125000) = 4605$ and $\pi(125000) = 11734$. Their ratio equals 0.39245, in good agreement with (355). The *index of irregularity* $j(p)$ is the number of B_n in Definition 40 divisible by p; regular primes have $j(p) = 0$ and irregular primes have $j(p) \geqq 1$. A related conjecture is

Conjecture 23.

$$J(N) = \sum_{p=3}^{N} j(p) \sim \frac{1}{2}\pi(N). \tag{356}$$

The heuristic argument is now even simpler if the same equidistribution is assumed. Wagstaff's data gives $J(125000) = 5842$ and $J(125000)/\pi(125000) = 0.49787$, in good agreement with (356). More to the point is the fact that $N = 125000$ is not exceptional: $J(N)/\pi(N)$ and $I(N)/\pi(N)$ both have only small fluctuations up to this limit.

Of the three conjectures, Conjectures 16, 22, and 23, the last is the weakest, but conceivably it may be the least difficult to prove. If it is proved, then Conjecture 16 is true for at least one-half of all prime exponents.

Turning to Conjecture 17, it is now true for all $p < 3 \cdot 10^9$ since, as we indicated above [91], the only violations of Wieferich's (208) for $p < 3 \cdot 10^9$ remain the old cases, $p = 1093$ and 3511. Prior to [91], everyone quoted the Lehmers' smaller bound 253, 747, 889, but this may have become invalid shortly after they computed it [27] in 1941. The reason is that they not only assumed the validity of the criteria in (208) and (209) but also that of all such criteria:

$$p^2 \nmid q^{p-1} - 1 \tag{357}$$

for every prime $q \leqq 43$. In 1948 Gunderson [102] questioned the validity of the proofs that had been given for (357) for the last three cases: $q = 37$, 41, and 43.

Nonetheless, using (357) only for $q \leqq 31$, he deduced a bound for Conjecture 17 that was larger than 253,747,889, namely

$$p < 1.1 \cdot 10^9.$$

He showed (Theorem N) that if

$$p^2 | q_i^{p-1} - 1 \tag{358}$$

for the first n primes: $q_1 = 2, q_2 = 3, q_3 = 5, \ldots, q_n$, then p satisfies the inequality

$$\frac{(2n-2)!}{(n-1)!\,(n-1)!}\,\frac{2}{n!}\,\frac{\left(\log \dfrac{p}{\sqrt{2}}\right)^n}{\log q_1 \cdot \log q_2 \cdots \log q_n} \leqq \frac{p-1}{2}. \tag{359}$$

Designating the left side by $f_n(p)$, one finds that the iterative sequence

$$p = 2f_n(p) + 1 \tag{360}$$

converges fairly rapidly to the desired bound for p. Since $31 = q_{11}$, the use of (360) for $n = 11$ gives Gunderson's bound for Conjecture 17 more precisely, namely,

$$p < 1{,}110{,}061{,}000.$$

If the validity of (357) is proved for $q_{12} = 37$ one gets a new bound:

$$p < 4{,}343{,}289{,}000. \tag{361}$$

If $q_{13} = 41$ and $q_{14} = 43$ are also good, this becomes

$$p < 57{,}441{,}749{,}000,$$

and if $q_{15} = 47, \ldots, q_{20} = 71$ are also good, we have

$$p < 32{,}905{,}961{,}000{,}000.$$

Since (361) is already better than the present bound $3 \cdot 10^9$, the order of the day seems clear: investigate (357) for $q = 37$. If it is true, then we have a new bound; if not, there must be an interesting mathematical reason for this failure.

Concerning Euler's generalization designated as Open Question 2 on page 158, I am pleased with my intuition there. I refused to call it a conjecture, since I said that there was no serious evidence for it. Several years later a counter-example

$$144^5 = 27^5 + 84^5 + 110^5 + 133^5 \tag{362}$$

was found by Lander and Parkin [103]. Curiously, no other counter-example is known. The most probable reason is that further computations, as in [104], have simply not gone far enough.

Since Open Question 2 is settled, let us replace it with

Open Question 3. *Is there a nontrivial solution of*

$$A^4 = B^4 + C^4 + D^4? \tag{363}$$

Although (363) has been investigated frequently, there is insufficient evidence to warrant a conjecture. One often reads that the methods of algebraic geometry are very powerful. Perhaps it is not too unfair to challenge the algebraic geometers with (363): find a solution or prove that none exists. No doubt algebraic geometry itself would be the main beneficiary, since new developments would probably be required.

Exercise 168 (W. Johnson [105]). Determine the probability of $j(p) = n$, using the previous assumption. For $n = 0$, we gave $P = e^{-1/2}$ above.

Exercise 169. The absence of Wieferich Squares p^2 for $3511 < p < 3 \cdot 10^9$ does not contradict Exercise 158, since the probable number in this interval is only 0.983. Using (208) and (209) and the sum $\sum_{p=3\cdot 10^9}^{\infty} p^{-2}$, what is the probability of a counter-example for Conjecture 17?

71. Binary Quadratic Forms with Negative Discriminants

The most classical of classical number theory is the theory of binary quadratic forms. Yet even here there has been significant development. We cannot adequately treat all of these topics here, since we largely confined ourselves above to the classical problems

$$p = a^2 + Nb^2, \qquad x^2 - Ny^2 = \pm 1$$

that initiated the subject and to their immediate generalizations.

Starting with Fermat's Theorem 60, we might add a survey of computational methods [106] and one new short-cut [107]. For Theorem 69, let us extend the data for $R(N)$ given in Ex. 119 with the results given in [108] and the references cited there:

$$R(10^6) = 3141549, \qquad R(10^8) = 314159053,$$

$$R(10^{10}) = 31415925457, \qquad R(10^{12}) = 3141592649625,$$

$$R(10^{14}) = 314159265350589.$$

There has also been interest in Landau's function $B(x)$, which counts each integer $n = a^2 + b^2 \leqq x$ only once no matter how many representations it may have [109].

In the generalization

$$rp = a^2 + Nb^2 \tag{364}$$

on page 167, we wish to make $r = 1$ if possible and to minimize it otherwise. This relates, as we indicated on pages 153, 154, and 168, to questions involving unique factorization and to those concerning the density of primes generated by quadratic polynomials. In an important development, H. M. Stark proved [110] that for negative N the quadratic field $k(\sqrt{N}\,)$ has unique factorization only for

$$N = -1, -2, -3, -7, -11, -19, -43, -67, -163. \tag{365}$$

A. Baker [111] and K. Heegner [112] have given other approaches to this long-sought theorem. Correspondingly, the famous polynomial $n^2 + n + 41$, which has -163 for its discriminant, must have a very high density of primes. In [56] we find that we should take $C = 3.31977$ in (338) with $k = 1$. Paul T. Rygg [113] has counted these primes up to $n = 10^6$, and his count does agree very well with (338).

For computational developments on (364) we refer to published tables such as [114], to reviews thereof, such as [115], and to improved algorithms, such as [83, sec. 6]. An example in the latter solves (364) for every N from 1 to 150 inclusive for a remarkable prime

$$p = 26437680473689 \tag{366}$$

that we will refer to repeatedly below. Such solutions are possible only because $(-N|p) = +1$ for all N between 1 and 150 for this prime. The generalization of Landau's $B(x)$ to $n = a^2 + Nb^2 \leqq x$ has been studied in [116].

Much (but not all) of the recent development in factorization methods involves binary quadratic forms either explicitly or implicitly. Our Theorem 76 above is closely related to the Lehmers' algorithm [97], which may be used both for factoring and for primality tests. The previously cited [78] has these same features; however, it derives its greater efficiency not from Theorem 76 but from more advanced ideas involving class groups and composition that we did not study above. We must therefore drop the topic, even though it would fit in nicely with our previous text; the class groups are Abelian and their cycle graphs are particularly informative. We continue with other references for new factorization methods in the next section.

In view of the historical importance of Pythagorean numbers (see Fermat's statement to Frenicle, on page 161) it is curious that the obvious three-dimensional analogue was not examined earlier. As far as I know, it is new. In how many ways can we solve

$$p^2 = a^2 + b^2 + c^2 \qquad \text{for} \qquad 0 < a \leqq b \leqq c \tag{367}$$

if p is an odd prime? The elegant answer is in

Theorem 86. *Write p uniquely as $p = 8n \pm 1$ or $p = 8n \pm 5$. Then* (367) *has exactly n solutions.*

EXAMPLES: $p = 3$, 7, and 13 each have one solution, $p = 5$ has none, and $p = 19$ has three.

$$9 = 1 + 4 + 4, \qquad 49 = 4 + 9 + 36, \qquad 169 = 9 + 16 + 144.$$
$$361 = 1 + 36 + 324 = 36 + 36 + 289 = 36 + 100 + 225.$$

For a proof, see [117]. It is based on known classical results involving ternary forms (see page 246).

EXERCISE 170. Determine the nine solutions of (367) for $p = 73$. Note that $73^2 - 12^2 = 61 \cdot 85$ gives rise to four of them.

72. Binary Quadratic Forms with Positive Discriminants

In contrast with (365) we list

Conjecture 24. *There are infinitely many quadratic fields $k(\sqrt{N}\,)$ for $N > 0$ that have unique factorization.*

This is an important conjecture, since its proof will require a deep insight not now available. For the large p in (366), $k(\sqrt{p}\,)$ has class number $h = 1$ and therefore unique factorization. Empirically, that is not surprising; for about 80% of known $k(\sqrt{p}\,)$, where p is a prime $\equiv 1 \pmod 4$, we have $h = 1$ [118]–[120], and this empirical density decreases only very slowly as p increases [121]. Therefore, the a-priori odds actually favor $h = 1$ for the prime in (366). While there are only nine cases in (365), many thousands of such fields have been recorded for $N > 0$.

The difference arises from the fact (page 173) that one has infinitely many units when $N > 0$. We must generalize Fermat's equation (236) to include the possibilities indicated in Exercises 124 and 144. That done, we have

Definition 45. If T and U are the smallest positive integers that satisfy

$$T^2 - U^2N = \pm 4 \tag{368}$$

then

$$\epsilon = (T + U\sqrt{N})/2 \tag{369}$$

is called the *fundamental unit* of $k(\sqrt{N})$ and $R = \log \epsilon$ is called its *regulator*.

To be brief, it is known that the product $2Rh/\sqrt{N}$ plays the same role if $N > 0$ that the product $\pi h/\sqrt{-N}$ does if $N < 0$. In the latter case, the class number grows (on the average) proportionally to $\sqrt{-N}$; while in the former, Rh grows (on the average) proportionally to $\sqrt{N}$. Thus, if R is big enough, there is no reason why h cannot equal 1 no matter how big N becomes. So the real question is this: Why are the fundamental units (369) frequently so large? This takes us back to the very beginning. When Fermat and Frenicle challenged the English (page 172) with $N = 61, 109, 151, 313, \ldots$, they may not have realized it, but $k(\sqrt{N})$ has $h = 1$ in all of those cases.

For the p of (366) the smallest u that satisfies $u^2 - pv^2 = -1$ has 9440605 decimal digits [122]. That makes even the answer to the famous Archimedes Cattle Problem [123], [124] look small. The regulator of $k(\sqrt{p})$ is 21737796.4. It is that large because (a) the class number is 1, (b) p is large, and (c) $(p|q) = +1$ for $q = 3, 5, 7, \ldots, 149$. This last point is significant, since an "average" p this size, not having this unusual property, would have a u with only 1116519 digits.

Digressing briefly, it is the last point (c) that gives p its mathematical interest (not its gigantic u). It is the smallest prime $\equiv 1 \pmod 8$ that has $(p|q) = +1$ for $q = 3$ to 149. The Riemann Hypothesis puts a limit on how long a run of residues a prime of a given size can have, and p was computed by the Lehmers and myself precisely to test this limit [125].

Had Frenicle persuaded Lord Brouncker to compute the continued fraction for $\sqrt{p}$, they would have found [126] that its period is 18331889. But a new development makes it possible to compute R accurately in a few seconds of machine time. Exercise 141 shows how to use symmetry to cut the computation in half. It turns out (surprisingly) that symmetry is not essential here; the use of composition and quadratic forms allow a doubling operation *anywhere* in the period, and therefore repeated doubling is also possible [122].

For $h = 1$ in cubic and quartic fields, see [120] and [127]–[129], while for three interesting continued fractions, see [130]–[132].

Returning to factorization, the continued-fraction method [133] is complicated but extremely powerful. An interpretation of it in terms of quadratic forms [134] is of interest; and subsequently this led to a greatly simplified method [135], [136], which loses much of the previous power but all of the complexity. It is now so simple and requires so little storage that one can factor

$$2^{60} + 2^{30} - 1 = 139001459 \cdot 8294312261,$$
$$2^{51} - 7 = 17174671 \cdot 131111671$$

on a little HP-67 even though this only computes with 10-digit numbers. Other recent developments in factorization are by Lehman [137], Pollard [138], and J. C. P. Miller [139]. For a survey article, see Guy [140].

EXERCISE 171. Since 17174671 has a unique representation $A^2 + 190B^2$ for $A = 3991$ and $B = 81$, it is prime. Why do we select 190? [141].

73. LUCAS AND PYTHAGORAS

Our estimate on page 198 that there will be "about 5" new prime M_p for $5000 < p < 50000$ needs little revision, if any. Four have been found for $p < 21000$ and "about 5" still seems a reasonable guess. Gillies [142] has found prime M_p for $p = 9689$, 9941 and 11213, and Tuckerman [143] has found M_{19937}.

Gillies included a statistical theory, based upon unproved hypotheses, which implies that about six or seven prime M_p should be expected in each decade: $A < p < 10A$. Ehrman studied these Gillies hypotheses [144] and interpreted previous data [145] on the distribution of the number of divisors M_p has below a given bound B. These distributions, and those in (331), constitute first steps in understanding M_p.

There has been no computation of M_p to my knowledge since that of Tuckerman [143]. That is surprising, since it was at that very time that Knuth had begun to publicize [146] the new Strassen-Schönhage "fast Fourier multiplication" algorithm for which one has

Theorem S. *It is possible to multiply two n-bit numbers in* $O(n \log n \log \log n)$ *steps.*

This leaves open the pertinent question: For what n does this become competitive with the older $O(n^2)$ multiplication? It does seem to offer an escape from our statement on page 195 that the Lucas arithmetic for M_p is roughly proportional to p^3, and I do not know why this has not been exploited.

We should add that the theory of Lucas sequences plays a large role in many of the new primality tests referenced above, not merely in tests for M_p.

Returning to the beginning of Chapter III, the Case for Pythagoreanism remains an important philosophical proposition. I know of no serious discussion or refutation that has appeared anywhere; Eves [147] merely copied our list without advancing the question. It is therefore unnecessary to strengthen the case here, but two additions and one subtraction should be made. The genetic code in DNA and recent theories of elementary particles are almost pure Pythagoreanism, and it is hard to conceive of two more fundamental things in the universe. On the other hand, let us delete Eddington's speculation that $hc/2\pi e^2 = 137$ from our page 137. It mars a good case, since subsequent measurements [148] have given

$$hc/2\pi e^2 = 137.0388 \pm 0.0019.$$

74. The Progress Report Concluded

We are nearly done; but even the Supplement and the commentary in the first edition References need updating. For more on Exercises 4S and 8S, see [149] and [150], respectively. Finite geometries, as in Exercise 5S, arise in interesting number-theoretic situations; cf. [120, page 30].

Waring's Conjecture (page 211) that every positive integer is the sum of $I(k)$ non-negative k-th powers is now even more "nearly" proven—but still not completely. Rosemarie Stemmler [151] completed a verification for all k up to 200,000, excluding the two hard cases $k = 4$ and 5. Mahler [152] had already shown that $g(k) \neq I(k)$ for at most a finite number of k. Continuing developments of Baker's method [111] suggest that a proof will be found for all $k > 200{,}000$, but this has not yet been done. As we indicated on page 212, $k = 4$ is the hardest case. Chen [153] has now proved that $g(5) = I(5) = 37$ and, while there has been progress on $g(4)$, its value remains unsettled. It seems likely that Waring's Conjecture will be completely proved in due course.

Dickson's valuable History [1] has been reprinted by Chelsea; and the dedicated scholar we called for on page 243 has turned out to be Wm. J. LeVeque. His six-volume [154] collection of reviews, while not quite equivalent to Dickson's *History*, is certainly a valuable aid to research.

This progress report confirms the statement in the 1962 preface that "number theory is very much a live subject." Even within the limited confines of our previous subject matter, the progress made since then is impressive.

STATEMENT ON FUNDAMENTALS

The logical starting point for a theory of the integers is Peano's five axioms. From these one can define addition and multiplication and prove all the fundamental laws of arithmetic, such as

$$\begin{aligned} a + b &= b + a, \\ a(b + c) &= ab + ac, \\ a(bc) &= (ab)c, \end{aligned}$$

etc. The reader knows that we have not done this. We have assumed all these fundamentals without proof, and even without explicit statement. Sometime, however, if he has not already done so, the reader should go through this development, and he can hardly do better than to read Landau's *Foundations of Analysis*, Chelsea, 1951.

Similarly we have skipped over the simpler properties of divisibility. We have not defined "divisor," "divisible," "even," etc. If there is an integer c such that

$$ac = b$$

we say a is a *divisor* of b. If 2 is a divisor of b we say b is *even*, etc. For these elementary definitions, and for such theorems as

$$a|b \text{ and } b|c \text{ implies } a|c,$$

$$a|b \text{ and } a|c \text{ implies } a|b + c,$$

etc., the reader is referred to Chapter I of a second book of Landau, *Elementary Number Theory*, Chelsea, 1958.

One of these elementary theorems should, however, be singled out for special mention. This is the *Division Algorithm*:

Theorem. *If* $a > 0$, *then for every* b *there are unique integers* q *and* r, *with* $0 \leqq r < a$, *such that*

$$b = qa + r;$$

that is, there is a unique quotient q *and a unique remainder* r.

This theorem is indeed a fundamental one in the theory of divisibility. It enters the theory via the Euclid Algorithm (page 8) and elsewhere.

The proof runs as follows. Let $b - x_1 a$ be the *smallest* non-negative integer of all integers of the form $b - xa$, with x an integer. Set $q = x_1$ and $r = b - qa$. Then one shows that $0 \leqq r < a$, whereas for any other x_2, $r_2 = b - x_2 a$ would not satisfy one, or the other, of the inequalities in $0 \leqq r_2 < a$.

The key argument is the fact that there exists a smallest non-negative $b - xa$. This is guaranteed by the Well-Ordering Principle (page 149). As is stated on page 149 the latter is equivalent to the principle of induction (Peano's fifth axiom), and thus is the principle which gives the integers their special (discrete) characteristics.

TABLE OF DEFINITIONS

Definition	*Page*
1) perfect number	1
2) prime, composite	3
3) greatest common divisor, (a, b), a is prime to b	5
4) $a \mid b$, $a \nmid b$	5
5) M_n, Mersenne number	14
6) $[x]$	14
7) $\pi(n)$	15
8) asymptotic, $f(x) \sim g(x)$	15
9) $\phi(n)$, Euler's phi function	20
10) $\pi_{a,b}(n)$	21
11) equinumerous	31
12) Legendre Symbol, $\left(\frac{a}{q}\right)$	33
13) Original Legendre Symbol, $(a \mid q)$	41
14) $P_a(N)$	48
15) $b \equiv c \pmod{a}$, congruent to, modulo, residue of, residue class, modulus	55
16) $a^{-1} \pmod{m}$, reciprocal	60
17) group, closed, associative, identity, inverse	60
18) Abelian group, finite group, order of the group	61
19) modulo multiplication group, $\mathfrak{M}_m$	61
20) quadratic residue, quadratic nonresidue	63
21) $\sqrt{a} \pmod{p}$	64
22) a is of order e modulo m	71
23) primitive root	73
24) cyclic group, generator	73
25) isomorphic groups, $a \longleftrightarrow b$, $\mathfrak{M}_m \rightleftarrows \mathfrak{M}_{m'}$	74
26) circular parity switch	76
27) cycle of a modulo m	83
28) subgroup	83
29) ϕ_m	92
30) Φ_m, characteristic factor, f_r	94
31) quadratfrei	114
32) fermatian	115
33) Fermat number, F_m	115

REFERENCES

The best single source for the historical aspects of number theory is, of course, the monumental

1. LEONARD EUGENE DICKSON, *History of the Theory of Numbers*, Chelsea, New York, 1971, Volumes I, II, III.

Our account here of the origin of Fermat's Theorem, of Cataldi's table of primes, etc. is largely drawn from this source. It would be highly desirable if a dedicated scholar, with access to appropriate funds and graduate assistants, should undertake to bring this history up-to-date.

We will not attempt, in this section, to indicate the source of everything in the present book, but will concentrate on the more modern theorems, conjectures, and tables, and on a few historically interesting points. The references will be numbered and listed under appropriate section headings.

SECTION 5

The primes, out to greater than 10^8, have been recently tabulated on microcards in

2. C. L. BAKER & F. J. GRUENBERGER, *The First Six Million Prime Numbers*, Madison, 1959.

For the last two entries in the table on page 15, see

3. D. H. LEHMER, "On the exact number of primes less than a given limit," *Ill. J. of Math.*, 3, (1959) p. 381–388.

SECTION 6

4. D. H. LEHMER, "Tables concerning the distribution of primes up to 37 millions" 1957, reviewed in *MTAC*, 13, (1959) p. 56–57.
5. M. KRAITCHIK, "Les Grand Nombres Premiers," *Sphinx*, Mar. 1938, p. 86.

For a complete account of the early work on Theorem 9 see

6. E. LANDAU, *Handbuch der Lehre von der Verteilung der Primzahlen*, Chelsea, New York, 1974, Chapters 1 and 2.

SECTION 8

For a table of $\phi(n)$ see

7. J. W. L. GLAISHER, *Number-Divisor Tables*, Cambridge, 1940, Table 1.

There have been numerous tables of $\pi_{a,b}(n)$. For discussion and further references, see

8. DANIEL SHANKS, "Quadratic residues and the distribution of primes," *MTAC*, 13, (1959) p. 272–284.

SECTION 11

Historically, Theorem 18 required a long time to get proved—analogous to the delayed proof in our treatment. It was first proven in its entirety by Lagrange in 1775—3 years *after* Euler determined the primality of M_{31}. But, of course, this proof did not use Gauss's Criterion, as we do on page 40. It does use Euler's Criterion, and theorems on the prime divisors of binary quadratic forms similar to Theorems 72 and 74 on page 166. See reference 12 below, page 209 for an account. By these latter-type theorems, Theorem 19 may follow directly, and not, as we show here, as a consequence of the harder Theorem 18. Thus if $q \mid 2\, M_p = N^2 - 2$, q must be of the form $q = x^2 - 2y^2$, and it follows at once that $q = 8k \pm 1$. See, in this connection, the remark on page 143 concerning the fact that quadratic residues arise most obviously in connection with binary quadratic forms.

SECTION 12

The large composite M_p on page 29 were obtained as a byproduct of the studies in reference 16. See Exercise 17 for the connection.

The two smallest pairs of twin primes $>10^{12}$ on page 30 are from reference 5, while the two largest pairs $<2^{47}$ are from unpublished work (June, 1961) by P. M. Fitzpatrick et al.

Conjectures 7 and 12 are from

9. G. H. HARDY & J. E. LITTLEWOOD, "Partitio numerorum III: On the expression of a number as a sum of primes," *Acta. Math.*, 44, (1923) p. 48.

The Goldbach Conjecture must surely strike a thoughtful reader as exceedingly curious. One *multiplies* primes; why add them? When Euler was attempting to prove Theorems 60 and 61 (pages 142, 143) in the 1740's, he was in frequent correspondence with Goldbach. The latter presumably conceived the idea that *if* every integer of the form $4k + 2$ is the sum of two primes of the form $4m + 1$, (a variant of the Conjecture), then Theorem 61 would follow for all integers of the form $4k + 2$ from Theorem 60. From this conclusion it easily follows that Theorem 61 is true for all positive integers. Empirically, Goldbach's notion seemed valid, at least if one allows 1 to be considered a prime. This was the accepted convention then. Thus $14 = 1 + 13 = 0^2 + 1^2 + 2^2 + 3^2$.

For interesting modern work in additive number theory arising out of Goldbach's Conjecture, see

10. KHINCHIN, *Three Pearls of Number Theory*, Graylock, Rochester, 1952, Chapter II.

For the much more difficult Vinogradoff theorem that every sufficiently large odd number is the sum of three primes, see

11. T. ESTERMANN, *Introduction to Modern Prime Number Theory*, Cambridge, 1952, Chapter 3.

SECTION 18

For the Original Legendre Symbol, see

12. ADRIEN-MARIE LEGENDRE, *Théorie des Nombres, Tome I*, Blanchard, Paris, 1955, p. 197.

SECTION 19

Until rather recently textbook proofs of Theorem 27 were distinctly more complex. The first publication of the original Frobenius proof in a book written in English appears to be in

13. WM. J. LEVEQUE, *Topics in Number Theory, Vol. I*, Addison-Wesley, Reading, 1956, p. 70.

Prior to this it had appeared in the next reference, being attributed to Frobenius, but without a detailed reference.

14. H. HASSE, *Vorlesungen über Zahlentheorie*, Springer, Berlin, 1950, p. 93.

The present author had not yet seen reference 14 in 1952 when he independently found the same proof.

15. DANIEL SHANKS, "The quadratic reciprocity law," Abstract 336 *t*, *Bull. A.M.S.*, 58, (1952), p. 452.

SECTION 20

Conjecture 12_1 is a rewording of a proposition in reference 9. For a more recent discussion and pertinent data, see

16. DANIEL SHANKS, "On the conjecture of Hardy & Littlewood concerning the number of primes of the form $n^2 + a$," *Math. Comp.*, 14, (1960), p. 321–332.

SECTION 30

17. DANIEL SHANKS, "A circular parity switch and applications to number theory," Abstract 543–7. *Notices*, Amer. Math. Soc., 5, (1958), p. 96.

SECTION 31

For recent Fermat composites and further references, see

18. G. A. PAXSON, "The compositeness of the thirteenth Fermat number," *Math. Comp.*, 15, (1961), p. 420.

SECTION 32

See reference 14, p. 68–69, and

19. A. J. C. CUNNINGHAM, "On the number of primes of the same residuacity," *Proc. London Math. Soc.*, ser. 2, 13, (1914), p. 258–272.
20. HERBERT BILHARZ, "Primdivisoren mit vorgegebener Primitivwurzel," *Math. Ann.*, 114, (1937), p. 476–492.

21. John W. Wrench, Jr., "Evaluation of Artin's constant and the twin-prime constant," *Math. Comp.*, 15, (1961), p. 396–398.

Section 33

The cycle graphs were investigated by the author in the early 1950's, but have not been previously published.

Section 39

22. Sidney Kravitz, "The congruence $2^{p-1} \equiv 1 \pmod{p^2}$ for $p < 100{,}000$," *Math. Comp.*, 14, (1960), p. 378.
23. P. Poulet, "Table des nombres composés vérifant le théorème de Fermat pour le module 2 jusqu'à 100,000,000," *Sphinx*, Mar. 1938, p. 42–51.
24. D. H. Lehmer, "On the converse of Fermat's Theorem, II," *Amer. Math. Monthly*, 56, (1949), p. 300–309.

Section 47

On pages 142–143 we indicate the historical background of $x^2 + Ny^2$ for $N = 1$ and $N = -2$. The case $N = +2$ is also interesting. Euler (and Goldbach) also attempted Theorem 61 as follows. If one can show that every integer of the form $8k + 3$ is the sum of three squares,

$$8k + 3 = (2a + 1)^2 + (2b + 1)^2 + (2c + 1)^2,$$

then that proves its equivalence, a theorem of Fermat which states that every integer is the sum of three triangular numbers:

$$k = \frac{a(a + 1)}{2} + \frac{b(b + 1)}{2} + \frac{c(c + 1)}{2}.$$

Further, one deduces

$$4k + 2 = (a - b)^2 + (a + b + 1)^2 + c^2 + (c + 1)^2,$$

so that Theorem 61 is valid for each such integer. As we indicated above, this implies its validity for all integers.

Now, by Theorem 72, all *primes* of the form $8k + 3$ can be represented by the binary form

$$8k + 3 = (2a + 1)^2 + 2(2b + 1)^2,$$

and this may be easily extended to all integers of the form $8k + 3$ that have prime divisors *only* of the form $8k + 1$ and $8k + 3$.

But this *binary* form does not suffice to obtain such representations as

$$8 \cdot 4 + 3 = 35 = 5 \cdot 7 = 1^2 + 3^2 + 5^2.$$

For these one needs *ternary* forms. The attempt therefore failed. Although

Euler did not prove Theorem 61 until Lagrange had already done so, his prolonged attack on this problem had several important byproducts.

SECTION 48

25. L. J. MORDELL, *Three Lectures on Fermat's Last Theorem*, Cambridge, 1921.

The remark concerning Bilharz on page 147 refers to reference 20.

SECTION 51

26. J. L. SELFRIDGE, C. A. NICHOL, & H. S. VANDIVER, "Proof of Fermat's Last Theorem for all prime exponents less than 4002," *Proc.* National Acad. Sci., 41, (1955), p. 970–973.

SECTION 52

See reference 25 and

27. D. H. & EMMA LEHMER, "On the first case of Fermat's Last Theorem," *Bull.* Amer. Math. Soc., 47, (1941), p. 139–142.

SECTION 63

28. D. H. LEHMER, "An extended theory of Lucas' functions," *Ann. of Math.*, 31, (1930), p. 419–448.

For the latest Mersenne primes and further references, see

29. ALEXANDER HURWITZ, "New Mersenne primes," *Math. Comp.*, 16, (1962), p. 249–251.

EXERCISE 34S.

For a survey on Waring's Conjecture, see

30. H. H. OSTMANN, *Additive Zahlentheorie, zweiter teil*, Springer, Berlin, 1956, p. 81–83.

For extensions to algebraic numbers, and the "easier" problem, see

31. C. L. SIEGEL, "Generalization of Waring's problem to algebraic number fields," *Amer. J. Math.*, 66, (1944), p. 122–136.
32. E. M. WRIGHT, "An easier Waring's problem," *Jour.*, London Math. Soc., 9, (1934), p. 267–272.
33. ROSEMARIE M. STEMMLER, "The easier Waring problem in algebraic number fields," *Acta. Arith.*, 6, (1961), p. 447–468.

For Linnik's famous, elementary, (but difficult) proof of Hilbert's Theorem (page 211) see reference 10, Chap. III.

EXERCISE 37S.

See reference 4. For a recent generalization of Conjectures 7, 11, 12, etc., see

34. PAUL T. BATEMAN & ROGER A. HORN, "A heuristic asymptotic formula concerning the distribution of prime numbers," *Math. Comp.*, 16, (1962), p. 363–367.

35. PETER HAGIS, JR., "A lower bound for the set of odd perfect numbers," *Math. Comp.*, 27, (1973), pp. 951–953.
36. BRYANT TUCKERMAN, "Review of Hagis UMT 53," *ibid*, pp. 1005–1006.
37. M. BUXTON and S. ELMORE, "An extension of lower bounds for odd perfect numbers, Abstract," *Notices*, Amer. Math. Soc., 23, (1976), p. A-55.
38. PETER HAGIS, JR., "Review of Tuckerman UMT 52," *Math. Comp.*, 27, (1973), pp. 1004–1005.
39. BRYANT TUCKERMAN, "A search procedure and lower bound for odd perfect numbers," *ibid*, pp. 943–949.
40. JAN BOHMAN, "On the number of primes less than a given limit," *BIT*, 12, (1972), pp. 576–577.
41. RICHARD P. BRENT, "The first occurrence of large gaps between successive primes," *Math. Comp.*, 27, (1973), pp. 959–963.
42. DANIEL SHANKS, "On maximal gaps between successive primes," *ibid*, 18, (1964), pp. 646–651.
43. RICHARD K. GUY and J. L. SELFRIDGE, "What drives an aliquot sequence?," *ibid*, 29, (1975), pp. 101–107.
44. DANIEL SHANKS and SIDNEY KRAVITZ, "On the distribution of Mersenne divisors," *ibid*, 21, (1967), pp. 97–101.
45. H. C. WILLIAMS and C. R. ZARNKE, "Some prime numbers of the forms $2A3^n + 1$ and $2A3^n - 1$," *ibid*, 26, (1972), pp. 995–998.
46. RICHARD P. BRENT, "Irregularities in the distribution of primes and twin primes," *ibid*, 29, (1975), pp. 43–56.
47. DANIEL SHANKS, "Review of Brent UMT 4," *ibid*, pp. 331.
48. DANIEL SHANKS, "Review of Brent UMT 21," *ibid*, 30, (1976), p. 379.
49. V. Brun, "La série $1/5 + 1/7 + 1/11 + 1/13 + 1/17 + 1/19 + 1/29 + 1/31 + 1/41 + 1/43 + 1/59 + 1/61 + \cdots$, où les dénominateurs sont 'nombres primes jumeaux' est convergente ou finie," *Bull. Sci. Math.*, 43, (1919), pp. 101–104, 124–128.
50. DANIEL SHANKS and JOHN W. WRENCH, JR., "Brun's constant," *Math. Comp.*, 28, (1974), pp. 293–299.
51. HANS RIESEL and GUNNAR GÖHL, "Some calculations related to Riemann's prime number formula," *ibid*, 24, (1970), pp. 969–983.
52. RICHARD P. BRENT, "The distribution of small gaps between successive primes," *ibid*, 28, (1974), pp. 315–324.
53. DANIEL SHANKS, "Supplementary data and remarks concerning a Hardy-Littlewood conjecture," *ibid*, 17, (1963), pp. 188–193.
54. M. C. WUNDERLICH, "On the Gaussian primes on the line $\mathrm{Im}(x) = 1$," *ibid*, 27, (1973), pp. 399–400.

55. H. DAVENPORT and A. SCHINZEL, "A note on certain arithmetical constants," *Illinois J. Math.*, 10, (1966), pp. 181–185.
56. DANIEL SHANKS, "Calculation and application of Epstein zeta functions," *Math. Comp.*, 29, (1975), pp. 271–287.
57. DOUGLAS HENSLEY and IAN RICHARDS, "On the incompatibility of two conjectures concerning primes," *Proc. Symposia Pure Math.*, 24, (1973), pp. 123–127. See also
57a. J. BARKLEY ROSSER and LOWELL SCHOENFELD, "Sharper bounds for the Chebyshev functions $\theta(x)$ and $\psi(x)$," *Math. Comp.*, 29, (1975), pp. 243–269.
58. M. L. STEIN and P. R. STEIN, "Experimental results on additive 2-bases," *ibid*, 19, (1965), pp. 427–434.
59. DANIEL SHANKS, "Review of Stein and Stein UMT 36," *ibid*, pp. 332–334.
60. JAN BOHMAN and CARL-ERIK FRÖBERG, "Numerical results on the Goldbach conjecture," *BIT*, 15, (1975), pp. 239–243.
61. H. HALBERSTAM and H-E. RICHERT, *Sieve Methods*, Academic Press, London, 1974.
62. J. CHEN, "On the representation of a large even integer as the sum of a prime and the product of at most two primes," *Sci. Sinica*, 16, (1973), pp. 157–176.
63. H. A. POGORZELSKI, "Goldbach Conjecture," *Crelle's J.*, 292, (1977), pp. 1–12.
64. A. SCHINZEL, "A remark on a paper of Bateman and Horn," *Math. Comp.*, 17, (1963), pp. 445–447.
65. DANIEL SHANKS, "A low density of primes," *Jour. Recreational Math.*, 5, (1971), pp. 272–275.
66. DANIEL SHANKS and MARVIN WUNDERLICH, "A low density of primes and the Primality Testing Package." (*To appear.*)
67. JOHN C. HALLYBURTON, JR. and JOHN BRILLHART, "Two new factors of Fermat numbers," *Math. Comp.*, 29, (1975), pp. 109–112.
68. G. MATTHEW and H. C. WILLIAMS, "Some new primes of the form $k \cdot 2^n + 1$," *ibid*, 31, (1977), pp. 797–798.
69. D. E. SHIPPEE, "Four new factors of Fermat numbers," *ibid*, 32, (1978).
70. D. H. LEHMER and EMMA LEHMER, "Heuristics anyone," *Studies in Mathematical Analysis and Related Topics*, Stanford Univ. Press, Stanford, 1962, pp. 202–210.
71. A. E. WESTERN and J. C. P. MILLER, *Indices and Primitive Roots*, Cambridge Univ. Press, Cambridge, 1968.

72. CHRISTOPHER HOOLEY, "On Artin's conjecture," *Crelle's J.*, 225, (1967), pp. 209–220.
73. DANIEL SHANKS, "Review of Baillie UMT 51," *Math. Comp.*, 29, (1975), pp. 1164–1165.
74. DANIEL SHANKS, "Fibonacci primitive roots," *Fibon. Quart.*, 10, (1972), pp. 163–168, 181.
75. H. W. LENSTRA, JR., "On Artin's conjecture and Euclid's algorithm in global fields," *Invent. Math.*, 42, (1977), pp. 201–224.
76. DANIEL SHANKS, "Review of Baillie UMT 6," *Math. Comp.*, v. 30, (1976), pp. 189–190.
77. DANIEL SHANKS, "Gauss's ternary form reduction and the 2-Sylow subgroup," *ibid*, 25, (1971), pp. 837–853.
78. DANIEL SHANKS, "Class number, a theory of factorization, and genera," *Proc. Symposia Pure Math.*, 20, (1970), pp. 415–440.
79. DANIEL SHANKS, "Dedekind zeta functions that have sums of Epstein zeta functions as factors." (*To appear.*)
80. DANIEL SHANKS, "Review of Hall and Senior, *The Groups of Order* 2^n $(n \leqq 6)$," Math. Comp., 19, (1965), *pp*. 335–337.
81. C. F. GAUSS, *Untersuchungen über Höhere Arithmetik*, Chelsea reprint edition, New York, 1965.
82. G. B. MATHEWS, *Theory of Numbers*, Chelsea reprint edition, New York, 1961.
83. DANIEL SHANKS, "Five number-theoretic algorithms," *Proc. Second Manitoba Conf. on Numerical Math.*, Winnipeg, (1972), pp. 51–70.
84. C. F. GAUSS, *Disquisitiones Arithmeticae*, Yale Univ. Press, New Haven, 1966.
84a. DANIEL SHANKS, "Review of the translation above," *Math. Comp.*, 20, (1966), pp. 617–618.
85. D. H. LEHMER, "Table errata 485 for P. Poulet," *ibid*, 25, (1971), pp. 944–945.
86. J. D. SWIFT, "Table errata 497 for P. Poulet," *ibid*, 26, (1972), p. 814.
87. J. L. SELFRIDGE and SAMUEL WAGSTAFF, JR., "Pseudoprimes, strong pseudoprimes, and Carmichael numbers," *ibid* (*To appear.*)
88. P. ERDÖS, "On almost primes," *Amer. Math. Monthly*, 57, (1950), pp. 404–407.
89. P. ERDÖS, "On pseudoprimes and Carmichael numbers," *Publ. Math. Debrecen*, 4, (1956), pp. 201–206.
90. A. SCHINZEL and W. SIERPIŃSKI, "Sur certaines hypothèses concernant les nombres premiers," *Acta Arith.*, 4, (1958), pp. 185–208.

91. J. Brillhart, J. Tonascia and P. Weinberger, "On the Fermat quotient," *Computers in Number Theory*, Academic Press, London, 1971, pp. 213–222.
92. Morris Newman, Daniel Shanks and H. C. Williams, "Simple groups of square order and an interesting sequence of primes." *Acta Arith.* (*To appear.*)
93. John Brillhart, D. H. Lehmer and J. L. Selfridge, "New primality criteria and factorizations of $2^m \pm 1$," *Math. Comp.*, 29, (1975), pp. 620–647.
94. H. C. Williams and J. S. Judd, "Determination of the primality of N by using factors of $N^2 \pm 1$," *ibid*, 30, (1976), pp. 157–172.
95. H. C. Williams and J. S. Judd, "Some algorithms for prime testing using generalized Lehmer functions," *ibid*, 30, (1976), pp. 867–886.
95a. H. C. Williams and R. Holte, "Some observations on primality testing," *ibid*, 32, (1978).
96. Gary L. Miller, "Riemann's hypothesis and tests for primality," *Jour. Computer and System Sci.*, 13, (1976), pp. 300–317.
97. D. H. Lehmer and Emma Lehmer, "A new factorization technique using quadratic forms," *Math. Comp.*, 28, (1974), pp. 625–635.
98. Henry B. Mann and Daniel Shanks, "A necessary and sufficient condition for primality, and its source," *Jour. Combinatorial Theory*, 13, (1972), pp. 131–134.
99. D. H. Lehmer, "Automation and pure mathematics," *Applications of Digital Computers*, Ginn, Boston, 1963, pp. 219–231.
100. C. L. Siegel, "Zu zwei Bemerkungen Kummers," *Nachr. Akad. Wiss. Göttingen*, (1964), No. 6, pp. 51–57.
101. Samuel S. Wagstaff, Jr., "The irregular primes to 125000," *Math. Comp.*, 32, (1978).
102. Norman G. Gunderson, *Derivation of Criteria for the First Case of Fermat's Last Theorem and the Combination of these Criteria to Produce a New Lower Bound for the Exponent*, Thesis, Cornell University, Sept. 1948.
103. L. J. Lander and T. R. Parkin, "A counterexample to Euler's sum of powers conjecture," *Math. Comp.*, 21, (1967), pp. 101–103.
104. L. J. Lander, T. R. Parkin and J. L. Selfridge, "A survey of equal sums of like powers," *ibid*, pp. 446–459.
105. Wells Johnson, "Irregular primes and cyclotomic invariants," *ibid*, 29, (1975), pp. 113–120.
106. Daniel Shanks, "Review of Diehl and Jordan UMT 19," *ibid*, 21, (1967), pp. 260–262. See also:
106a. Daniel Shanks, "Use of the infrastructure of a real quadratic

field in computing $p = A^2 + B^2$." (*To appear.*)

107. JOHN BRILLHART, "Note on representing a prime as a sum of two squares," *Math. Comp.*, 26, (1972), pp. 1011–1013.
108. W. C. MITCHELL, "The number of lattice points in a k-dimensional hypersphere," *ibid*, 20, (1966), pp. 300–310.
109. DANIEL SHANKS, "The second-order term in the asymptotic expansion of $B(x)$," *ibid*, 18, (1964), pp. 75–86.
110. H. M. STARK, "A complete determination of the complex quadratic fields of class-number one," *Michigan Math. J.*, 14, (1967), pp. 1–27.
111. A. BAKER, "Linear forms in the logarithms of algebraic numbers," *Mathematika*, 13, (1966), pp. 204–216.
112. K. HEEGNER, "Diophantische Analysis und Modulfunktionen," *Math. Zeit.*, 56, (1952), pp. 227–253.
113. DANIEL SHANKS, "Review of Rygg UMT," *Math. Comp.*, (*To appear.*)
114. H GUPTA, M. S. CHEMMA, A. MEHTA and O. P. GUPTA, *Representation of Primes by Quadratic Forms*, Royal Soc. Math. Tables No. 5, Cambridge Univ. Press, New York 1960.
115. DANIEL SHANKS, "Review of previous table," *Math. Comp.*, 15, (1961), pp. 83–84.
116. DANIEL SHANKS and LARRY P. SCHMID, "Variations on a theorem of Landau. Part I," *ibid*, 20, (1966), pp. 551–569.
117. DANIEL SHANKS, "Review of Forbes and Lal Table," *ibid*, 25, (1971), p. 630.
118. DANIEL SHANKS, "Review of Kloss, Newman and Ordman UMT 10," *ibid*, 23, (1969), pp. 213–214.
119. M. D. HENDY, "The distribution of ideal class numbers of real quadratic fields," *ibid*, 29, (1975), pp. 1129–1134.
120. DANIEL SHANKS, "A survey of quadratic, cubic and quartic algebraic number fields (from a computational point of view)," *Proc. Seventh Southeastern Conf. on Combinatorics, Graph Theory and Computing*, Louisiana State Univ., Baton Rouge, (1976), pp. 15–40.
121. RICHARD P. LAKEIN, "Review of Kuroda UMT 9," *Math. Comp.*, 29, (1975), pp. 335–336.
122. DANIEL SHANKS, "The infrastructure of a real quadratic field and its applications," *Proc. 1972 Number Theory Conf.*, Boulder, Colorado, (1972), pp. 217–224.
123. H. C. WILLIAMS, R. A. GERMAN and C. R. ZARNKE, "Solution of the cattle problem of Archimedes," *Math. Comp.*, 19, (1965), pp. 671–674.

124. Daniel Shanks, "Review of foregoing UMT 114," *ibid*, pp. 686–687.
125. Daniel Shanks, "Systematic examination of Littlewood's bounds on $L(1, \chi)$," *Proc. Symposia Pure Math.*, 24, (1973).
126. Daniel Shanks, "Review of Williams, Henderson and Wright UMT 11," *Math. Comp.*, 28, (1974), pp. 333–334.
127. Daniel Shanks, "Review of Barrucand, Williams and Baniuk UMT 20," *ibid*, 30, (1976), pp. 377–379.
128. H. C. Williams, "Certain pure cubic fields with class-number one," *ibid*, 31, (1977), pp. 578–580.
129. Richard B. Lakein, "Computation of the ideal class group of certain complex quartic fields. II," *ibid*, 29, (1975), pp. 137–144.
130. John W. Wrench, Jr. and Daniel Shanks, "Questions concerning Khintchine's constant and the efficient computation of regular continued fractions," *ibid*, 20, (1966), pp. 444–448.
131. Richard P. Brent, "Computation of the regular continued fraction for Euler's constant," *ibid*, 31, (1977), pp. 771–777.
132. John W. Wrench, Jr., "Review of Gosper Continued Fraction for π in UMT 15," *ibid*, 31, (1977), p. 1044.
133. Michael A. Morrison and John Brillhart, "A method of factoring and the factorization of F_7," *ibid*, 29, (1975), pp. 183–205.
134. Daniel Shanks, "Analysis and improvement of the continued fraction method of factorization," *Notices*, Amer. Math. Soc., 22, (1975), Abstract on A-68. A paper with the same title will appear.
135. Daniel Shanks, "Recent applications of the infrastructure of real quadratic fields $Q(\sqrt{N})$," *Notices*, Amer. Math. Soc., 23, (1976), Abstract on A-59.
136. Daniel Shanks, "Square-form factorization, a simple $O(N^{1/4})$ algorithm." (*To appear.*)
137. R. Sherman Lehman, "Factoring large integers," *Math. Comp.*, 28, (1974), pp. 637–646.
138. J. M. Pollard, "A Monte Carlo method for factorization," *BIT*, 15, (1975), pp. 331–334.
139. J. C. P. Miller, "On factorisation, with a suggested new approach," *Math. Comp.*, 29, (1975), pp. 155–172.
140. Richard K. Guy, "How to factor a number," *Proc. Fifth Manitoba Conf. on Numerical Math.*, Winnipeg, (1976), pp. 49–89.
141. Daniel Shanks, "Use of one class per genus in primality tests." (*To appear.*)
142. Donald B. Gillies, "Three new Mersenne primes and a statistical theory," *Math. Comp.*, 18, (1964), pp. 93–95.

143. BRYANT TUCKERMAN, "The 24th Mersenne prime," *Proc. Nat. Acad. Sci. USA*, 68, (1971), pp. 2319–2320.
144. JOHN R. EHRMAN, "The number of prime divisors of certain Mersenne numbers," *Math Comp.*, 21, (1967), pp. 700–704.
145. DANIEL SHANKS, "Review of Kravitz and Madachy UMT 113," *ibid*, 19, (1965), p. 113; Corrigenda, *ibid*, 27, (1973), p. 453.
146. DANIEL SHANKS, "Review of Knuth Report STAN-CS-71-194," *ibid*, 25, (1971), pp. 937–938.
147. HOWARD W. EVES, *In Mathematical Circles, Quadrants I and II*, Prindle, Weber and Schmidt, Boston, 1969, See 55°, 56°, 57° on pp. 36–37.
148. DANIEL SHANKS, "Review of McNish Report on Fundamental Constants," *Math. Comp.*, 18, (1964), p. 335.
149. DANIEL SHANKS, "A duality relation in exterior ballistics," *SIAM Rev.*, 6, (1964), pp. 54–56.
150. DANIEL SHANKS, "Review of Yates, *Prime Period Lengths*," *Math. Comp.*, 29, (1975), pp. 1162–1163.
151. ROSEMARIE M. STEMMLER, "The ideal Waring theorem for exponents 401–200,000," *ibid*, 18, (1964), pp. 144–146.
152. K. MAHLER, "On the fractional parts of the powers of a rational number (II)," *Mathematika*, 4, (1957), pp. 122–124.
153. JING-RUN CHEN, "Waring's problem for $g(5)$," *Sci. Sinica*, 13, (1964), pp. 1547–1568.
154. WILLIAM J. LEVEQUE (Editor), *Reviews in Number Theory*, Amer. Math. Soc., 1974.

INDEX